GW00720640

DEVELOPING A SYSTEMS VIEW OF EDUCATION

THE SYSTEMS-MODEL APPROACH

BELA H. BANATHY

LEAR SIEGLER, INC./FEARON PUBLISHERS
Belmont, California

To my wife, Eva

Library of Congress Catalog Card Number: 72-95008.

ISBN−O−8224−6700−3.

Printed in the United States of America.

P. W. J. Howard.

CONTENTS

PREFACE   v

DEVELOPING A SYSTEMS VIEW   1

    Mapping the Road: An Introduction to Three Systems
      Models   3

    Relevance, Rationality, and Renewal   4
    Expectations   4

THE SYSTEMS-ENVIRONMENT MODEL   5

    Section 2-1:   The General Systems-Environment
                     Model   6
    Section 2-2:   The School-Society Model   11
    Section 2-3:   Discussion and Application   14

THE STILL-PICTURE MODEL   19

    Section 3-1:   The General Still-Picture Model   20
    Section 3-2:   The Still-Picture Model of Schooling   26
    Section 3-3:   Discussion and Application   31

1

2

3

THE MOTION-PICTURE MODEL   35

       Section 4-1:   The General Motion-Picture Model   36
       Section 4-2:   The Transformation of the General
                    Motion-Picture Model   57
       Section 4-3:   Discussion and Application   70

THE ACTIVATION OF THE SYSTEMS VIEW   73

       Domains of Application   74
       Exploration   76
       Model Building   77
       Design, Development, Validation, and
         Implementation   81

GLOSSARY   85

4

5

PREFACE

The purpose of this book is to help you, the reader, acquire a systems view and learn how to apply it to the solution of problems. As you work with the systems concepts, principles, and models introduced in this book, applying them to your own field of interest, you will develop the ability to approach problems systemically.

In an experimental form, this book has been used in a variety of ways for several years. Individuals have used it as a self-directed learning resource. When used in this way it is important that the learner discuss his or her findings and solutions with someone who is familiar with systems thinking.

The book has also been used in seminars and applied to many subjects, such as instructional systems, educational technology, new models of schooling, library services, rapid transit, health services, social services, air transportation, flood control, and many others. Analyses of those systems, which were guided by the questions introduced in this book, led to ideas for designing or improving aspects of the systems under consideration. During the first half of the seminar, the participants worked with the first four chapters of this book. They used the models presented in those chapters to describe and analyze their sytsems. During the second half of the seminar, they used the systems models approach described in the last chapter to carry out a project of problem solution, design, or development.

The material presented in this book has also been used as a resource in research and development. It guided the conceptualization of an educational planning system and a resources development system for educational R & D. In addition, it has been used as a basis for the exploratory analysis and design of a career education model.

This book should be useful to anyone interested in developing an understanding of systems thinking. For those readers who are unfamiliar with systems terminology, a glossary of key terms has been provided at the end of the book.

B.H.B.

DEVELOPING
A
SYSTEMS VIEW

1

The systems view is a way of looking at ourselves, at the environment we live in, and at the entities that surround us or that we are part of. Having a description of systems theory or even an understanding of systems concepts and principles, however, does not constitute having a systems view. The systems view is a way of thinking, a point of view. And there is a logical way in which it can be developed.

By observing systems, studying their behavior, and exploring their descriptions, we can learn to recognize characteristics that are common to systems. As a result of this examination we will be able to define system concepts and discover certain systems principles.

When we have defined a set of concepts and discovered how concepts lead to principles, we will be in a position to look for relationships between principles. We will then organize these principles into certain conceptual schemes, which we will call *systems models.*

This process of starting from an observation and arriving at a model of a system constitutes the first phase of developing a systems view. The second phase is the "internalization" of the models—the integration of systems concepts, principles, and models into your own thinking. This kind of integration is the basis for a systems view. These two phases can be described in systems terms and shown in the form of a systems model (see Figure 1-1).

Phase One

- We gather information from observing various systems to provide *input* into our thinking.
- We *transform* this input into generalizations about systems.
- Then we produce an *output* of systems concepts and principles and of systems models built from these concepts and principles.
- We return to the first step, which was observation, to find out if the model we built agrees with reality. This process is called *feedback*. Using feedback, we can make any needed *adjustments*.

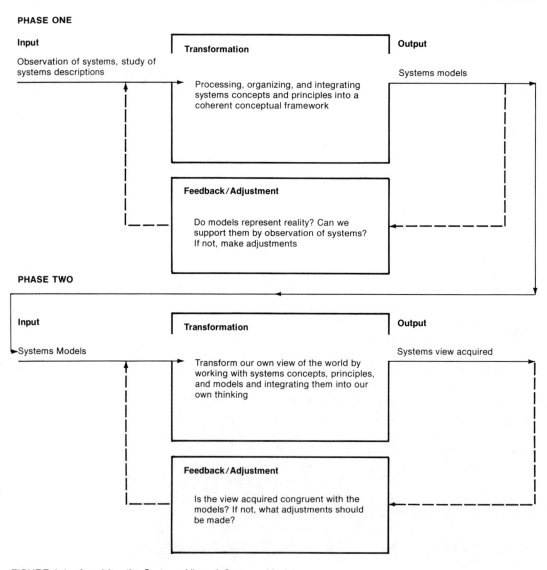

PHASE ONE

Input

Observation of systems, study of systems descriptions

Transformation

Processing, organizing, and integrating systems concepts and principles into a coherent conceptual framework

Output

Systems models

Feedback/Adjustment

Do models represent reality? Can we support them by observation of systems? If not, make adjustments

PHASE TWO

Input

Systems Models

Transformation

Transform our own view of the world by working with systems concepts, principles, and models and integrating them into our own thinking

Output

Systems view acquired

Feedback/Adjustment

Is the view acquired congruent with the models? If not, what adjustments should be made?

FIGURE 1·1   Acquiring the Systems View: A Systems Model

Phase Two

- For *input* we observe systems concepts, principles, and models.
- We *transform* our view of the world and ourselves, passing from an initial stage of nonsystems thinking to a point where we have integrated our observations into our own way of seeing the world. We arrive at the *output* state with a new way of thinking that enables us to view our own universe and everything in it in the systems way.
- Using *feedback*, we return to our models and check the adequacy of our systems view against them.

The transformation will come about from an acceptance of systems concepts, principles, and models and from their continuous application. Both acceptance and applications of these systems factors are required in developing our own systems view. Once we understand and accept these concepts, principles, and models, we still have to work with them for quite a while before we can develop a systems view and make it our own. Figure 1-1 demonstrates the processes of acquiring the systems view.

In the chapters that follow we will observe a set of systems models along with the concepts and principles that constitute these models.

MAPPING THE ROAD: AN INTRODUCTION TO THREE SYSTEMS MODELS

A model is a representation of reality. It is an expression of a mental image, an abstraction. This image is the product of our experience. It can be described, depicted, or displayed. The organized expression or description of this image is its model.

Models are useful as a frame of reference within which we can examine and talk about what the model represents. We work with models all the time. When we exchange ideas about something, we usually do so by using conceptual models. In a discussion, it is helpful to have a common model, or a common frame of reference, so that we have some assurance that everyone is talking about the same thing.

Because our purpose here is to talk about education as a system, it it important for us to create a common frame of reference for our discourse —to build models of systems and then systems models of education. Concepts and principles that are characteristic of systems can be organized into a systems model. These systems concepts and principles, then, can be organized into systems models of education.

In general systems research, we aim to develop general systems models that represent one or more classes of systems, and our present examination is limited to a single class of systems—the synthetic, man-made, living, social systems. We have selected this class as a focus because education is part of this kind of system. Once we have built a model from generalizations representing this class of systems, we will transfer these generalizations to education and thus build systems models of education.

There are many ways that systems concepts and principles can be built into models. It is useful to organize concepts and principles that are relevant to system into tree complementary models:

- One model examines systems in the context of their environment and organizes concepts and principles relevant to this examination. This model can be called the *systems-environment model* or the *systems-context model.*
- Another model focuses on what the system is, what it looks like, how it is organized. This model is a *spatial structural model.*
- The third model examines the behavior of the system over a period of time, and it tells us how the system operates. This model is a *process model.*

The motivation for constructing a model should be emphasized—models are useful only if they relate to or represent reality, if they help us to cope with a real situation, and if they can be used in solving problems.

There are two reasons for working with these three models: (1) their logic and (2) their usefulness to work in education. They provide a framework to use in finding and examining ways to solve problems in education.

RELEVANCE, RATIONALITY, AND RENEWAL

If we attempt to develop and study inventories of educational problems and synthesize them, we find that they form three major clusters that can be described in terms of our three proposed models: (1) the presence or absence of relevance, (2) rationality, and (3) educational renewal and improvement. More specifically, education faces the problems of its relevance to the learner and of meeting the multitudinous needs of society. Relevance probes into the ability of the schools to respond to the changing conditions, needs, and demands of the learner and of the society. The systems model that offers a conceptual framework for the examination of these and similar problems is the *systems-environment model.* The rationality of the way the school is organized and structured, and the cause-and-effect relationships between inputs and outcomes are questions that can be examined best within the *structural model.* Educational renewal and improvement are process questions which can be examined within the framework of the *behavioral* or *process model.*

EXPECTATIONS

What outcome can we expect from developing these three systems models? First we can use the models as bases upon which to develop corresponding systems models of education. Second, by working with these models, we will be able to develop a systems view in general and a systems view of education in particular. Finally, we will put our newly acquired systemic thinking to use by solving educational problems and designing new educational programs.

THE
SYSTEMS-ENVIRONMENT
MODEL

In this chapter, we will develop a systems-environment model of education. First, we will examine general systems concepts and principles and build a general systems-environment model. Then we will jointly transform this model into the context of society and schooling and thus build a systems-environment model of education. Finally, as a move toward a systems-environment view that applies to your own area of education, you will have the opportunity to analyze that area from the systems point of view. To accomplish these tasks, we will proceed in the following way: Study the description of the general systems-environment model presented in Section 2-1. In the empty space on the right-hand side of the pages, transform the general systems statements into the context of school-society relationships. Study the interpretation of a systems-environment model of education presented in Section 2-2 and compare it to your transformation. Then, develop a synthesis of the two transformations if you think that such a synthesis would be an improvement. Study the discussion in Section 2-3 and follow the guidelines for making an analysis of a specific aspect of education or schooling that you select. (*Education* as used here means formal education in a school.)

5

SECTION 2-1   THE GENERAL SYSTEMS-ENVIRONMENT MODEL

In this section we will identify some of the key concepts concerning the relationships of systems to their environment. We can then merge related concepts and derive principles from them—after which we will be able to organize the principles into a model. (In our terms, a *concept* can be "input"; a *principle* is essentially a law—for example: The more open the system is, the more kinds of input it has to cope with. A *model* is a theoretical scheme, such as a display of the various principles relevant to systems-environment interactions.)

As you read the description of general systems concepts, principles, and models, ask yourself these questions: How do these statements apply to formal education or schooling? How can the generalized statements be interpreted in a way that will reflect the relationships of the school to its environment? Avoid simply substituting generalized terms for terms that refer to the school-society concept. *Transform* the descriptions rather than translate them. Reformulate the ideas expressed in the description of general systems concepts in a way that will truly characterize relationships between the school and society.

THE GENERAL SYSTEMS-ENVIRONMENT MODEL

System Entities and Their Functions

Systems exist within a *given space* that is set aside from their environments by their boundaries.

The *environment* is the context within which a system exists. It is composed of all the things that surround the system, and it includes everything that may affect the system or that may be affected by the system.

The *boundaries* of a system delimit the system space and set aside from the environment all those entities, their attributes and relationships, that belong to the system. The boundaries of the types of systems we are examining are not usually drawn sharply and they do not isolate the system from its environment. There are breaks in the boundaries through which the system communicates and interacts with its environment, and through which the system receives its input and sends its output back into the environment.

The *input* is a collective term that refers to everything that the system receives from its

YOUR TRANSFORMATION

environment. The environment establishes the system—or the system establishes itself in an environment—in order to satisfy certain environmental expectations, demands, and requirements. These *expectations, demands,* and *requirements* make up some of the input, as do *constraints* under which the system is to operate. Furthermore, *resources,* such as information, people, energies, materials, and so forth, that come from the environment are also input.

Output, another collective term, is whatever the system sends back into its environment. It is this output that enables the system to meet the expectations, requirements, and demands of its environment.

Figure 2-1 represents an initial systems-environment model. It interrelates a few basic concepts and principles that we have discussed.

Relationships

The environment can also be thought of as a system, a larger one that includes the system under consideration. The larger system that surrounds the system can be called the *suprasystem.* The suprasystem may be made up of a set of systems; thus, it may also be thought of as a *multisystem.* Systems within a suprasystem are *peer systems.* Relationships among peer systems can be of three kinds: hierarchical, centralized, and equalitarian. A *hierarchical relationship* implies that one subsystem is subordinate to another. In a *centralized relationship,* one system plays a central role and other systems are related to it and are arranged

YOUR TRANSFORMATION

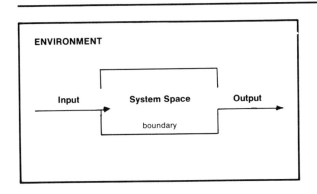

FIGURE 2·1   An Initial Systems-Environment Model

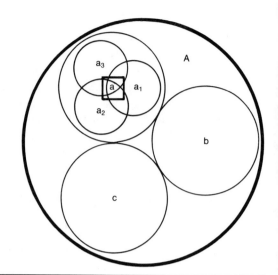

FIGURE 2·2 Systems Relationships. If we are examining a system *a*, then *A* is the suprasystem in which it exists, *b* and *c* are its peer systems, and a_1, a_2, and a_3 are its subsystems. If we select *A* as our primary level of examination, *a*, *b*, and *c* are subsystems.

around it. In an *equalitarian relationship,* none of the peer systems has a dominant or central role. Some systems can be structured down to *subsystems.* Subsystems can have the same kinds of relationships to each other that systems have. Figure 2-2 displays some systems relationships.

Further examination of the relationships between a system and its environment brings into focus another characteristic: the open or closed nature of systems. A system is *closed* if it is sealed off from its environment by its boundaries and if there is no trading between the system and its environment. A system is *open* if its boundaries are not well defined or have breaks that enable the system to interact with its environment. The larger these breaks are, the more open the system is and the more kinds of input and output it has to cope with. For practical purposes, we cannot speak of completely closed or completely open systems. Closedness and openness are matters of degree.

The more varied and complex the input, the more complex is the system and its output. The less varied the input, then the less varied the output, and the more closed the system is. A totally open system, on the other hand, implies totally unstable, unpredictable, and random input and output. This means the absence of specified and stated requirements,

YOUR TRANSFORMATION

demands, constraints, resources, and other factors that define what the system is. The types of systems with which we are concerned in education are definitely not closed, but neither are they fully open. They are systems that have to cope with a certain randomness of input but that still have a specific purpose and a specific expected output. In these systems there is some degree of *closure by control*—that is, an ability to adjust the breaks in the boundaries and to regulate input and output.

These systems can also *adjust* themselves to the changing requirements of their environment by modifying their own operations in order to produce the kind of output expected by the environment. We shall call these systems that are compatible with their environment *adaptive* and *self-regulating*.

System Control

In order for a system to survive, it needs to be adaptive. If the system ceases to produce the expected output and fails to remain acceptable to the environment, there are three options:

□ The system *adjusts* itself in order to produce the expected output.
□ The environment changes its expectation to *accommodate* the system.
□ The system is *terminated*.

Self-regulating adaptiveness is exercised by the system through its *feedback*. Feedback reintroduces information about the state of the output into the system. Adjustments can be made on the basis of this information. Through feedback, the *actual* state of the output is continuously compared with the *expected* state of the output. Feedback enables us to correct for differences between what is actually produced by the system and what the environment expects it to produce. Feedback-employing systems are adaptive and self-regulating, and, therefore, controlled systems. A controlled system state means an *ordered* state where the seemingly opposing drives of accommodation

YOUR TRANSFORMATION

and adjustment are harmonized as *system-environment coactions. System control* makes it possible for the system to adjust to its environment and, at the same time, it influences the environment so that it will accommodate the system. In this way, the environment maintains its responsiveness toward the system. The purpose of system control through feedback adjustment and environmental accommodation is to ensure a condition that will bring about an actual output that corresponds to the expected output. Now that we have taken feedback adjustment and environment response into account, we are in a position to complete the systems-environment model (see **Figure 2-3**).

Summary of the Systems-Environment Model

Systems exist in their environment, from which they are set apart by their boundaries. Some systems are rather closed and are isolated from their environment by their boundaries. However, at this time we are considering systems that are somewhat open, systems that have breaks in their boundaries enabling trade with their environment through input-output interactions. Systems of this kind are adaptive. They maintain compatibility by adjusting to the demands and expectations of their environment. This adjustment is made possible through self-regulating feedback control, which activates changes in order to ensure that the

YOUR TRANSFORMATION

FIGURE 2·3   The Systems-Environment Model

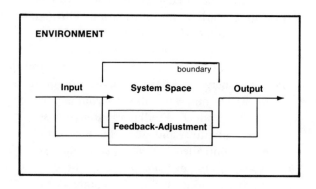

system output will be acceptable to the en-vironment. The environment, on the other hand, responds to and accommodates the needs of the system in order to facilitate the production of adequate output. Adjustment and accommodation, then, are the key sys-tems-environment coactions.

YOUR TRANSFORMATION

SECTION 2-2   THE SCHOOL-SOCIETY MODEL

In this section we will consider the transformation of the systems-environment model into education. We will call this transformation the school-society model. Compare this transformation with the one you de-veloped in Section 2-1. Then develop a synthesis of the two transforma-tions if you think that such synthesis would be an improvement. Record your synthesis on the right-hand sides of the pages.

SCHOOL AND SOCIETY AS SYSTEM AND ENVIRONMENT

YOUR SYNTHESIS

How the School Operates in Society

The *society* in general and a given community in particular is the context, or the environ-ment, within which the school exists. The society not only surrounds the school but creates it and is affected by its own creation.

The *boundaries* of the school can be de-scribed in physical, economic, psychobiolog-ical, psychological, and social terms. The phys-ical boundaries of the school are multiple. One such boundary is the limits of the phys-ical plant. A larger boundary is the geograph-ical area within which the school operates and which it serves.

Economic boundaries can be defined in terms of resources available to the school. Psychobiological boundaries refer to general student characteristics (such as age or apti-tudes). Psychological bounds refer to feelings, attitudes, and perceptions about the school.

The school receives multiple *input* from the society. The society establishes specific ex-pectations and *requirements* for the school.

However, the society also provides the school's *resources*—its human, financial, and material means and its facilities. Society sends pupils into the school to be educated. Students are the most important input and the key system entities of the school. The society also sets certain *constraints* within which the school is to operate. The pupils who have been educated and the knowledge that has been produced are the main *output* of the school into society. Figure 2-4 depicts some of these relationships.

Social, physical, psychological, and other kinds of boundaries define the space in which the school operates. The less defined these boundaries are, and the more breaks there are through which the school can interact with the society, the more open the school is and, consequently, the more kinds of input the school has to cope with.

Relationships Between the School and Society

Society is the *suprasystem* of the school, and the school is a subsystem of society. Other subsystems of society include government, industry, business, religion, and so on. Thus, society is a multisystem but so also is the school, which has numerous subsystems such as its instructional and administrative subsystems.

Government, industry, and business are among the *peer systems* of the school. The school interacts with these peer systems by influencing them and by being affected by them.

YOUR SYNTHESIS

FIGURE 2-4 An Initial School-Society Model. The greater the number of requirements the school is obliged to fulfill and the more variations there are in pupils and resources, the more complex the school becomes as a system.

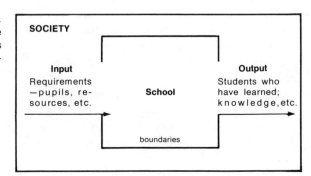

The peer relationships between the subsystems of the school are centralized (even though this may not be evident at times). If the purpose of education is learning, then the central subsystem is the one that is primarily responsible for facilitating learning. This would be the *instructional subsystem;* therefore, it should be the instructional subsystem around which other subsystems are built in the school.

System Control in the School

The school is an *open system,* but it is not completely open. Its inputs are regulated, its constraints identified, and its output specified. The school system's control is directed at the attainment of its output: the education of its pupils. Still, it is expected to cope with a certain degree of randomness and with continuous change in input: the society's demands, expectations, and constraints are often not specified clearly, and there are always great variations in individual differences among pupils.

A school in environment *A* will have to operate differently from one in environment *B*. (A school operating in the inner city of a metropolis, for example, needs to adapt itself to requirements that are quite different from those in rural areas.) The school should be able to adjust to differences in environmental demands and to produce the output specified by its environment. By adjusting, the school manifests its responsibility toward the society and maintains its compatibility with its environment.

Many questions have been raised about the school's ability to adapt to environmental changes and about its ability to bring about expected outcomes. If a school lacks accountability by not being able to guarantee educational outcomes, the society becomes increasingly reluctant to accept the school's demands for additional resources.

The output of education is continuously assessed and measured. On the basis of that output, adjustments can be designed. Feedback control begins with the process of collecting

YOUR SYNTHESIS

student performance data and comparing these data with the *performance model* of educational goals. The results of the comparison are analyzed, interpreted, and then fed back into the system for use in making adjustments in the system.

Through the use of its self-regulating, control-based feedback, the school can remain compatible with its environment, fulfill its expectations, and realize desired educational outcomes. In this case, the school can expect that the society will accommodate it and provide the resources and support it needs in order to be able to function.

Summary of the School-Society Model

Society is the environment of the school. From society, the school receives its input—that is, its human and material resources, expectations, and constraints. The school sends its output—the individual who has been educated and the knowledge produced—into the society. The school should satisfy environmental expectations and meet the changing requirements of its suprasystem, the society. Society, on the other hand, should accommodate the school by responding to its needs in order to facilitate the attainment of expected educational outcomes.

YOUR SYNTHESIS

SECTION 2-3   DISCUSSION AND APPLICATION

Discussion

The purpose of working with the system-environment model in the two preceding sections of this chapter was to foster the development of a systems view related to systems-environmental concerns: the environment we live in and the entities that surround us or that we are part of.

The development of a systems view is often blocked by certain predispositions, which, if they are present in our thinking, have to be changed. The recognition and understanding of such dispositions is important, for until we become aware of their existence, we cannot deal with them.

People generally prefer to operate in closed systems. Even when a system is open, we often tend to make it less so by narrowing the breaks in the system boundary in order to decrease the interaction and trade between the system and its environment. We often do this because we feel more secure when we operate on well-known ground surrounded by familiar landmarks.

The systems view places a high premium on the continuous and well-balanced interaction of the system and its environment. Systems exist to satisfy their environments, and once they lose their sensitivity to the needs of the environment, they begin to move toward termination. How to achieve a balanced interaction between school and society—system and environment—is a major challenge at all levels of education.

There are three specific ways that we can enhance the balance of adjustment and accommodation between the system and its environment. First, we must move out from our system space and become well acquainted with the environment. We must become aware of the environment's requirements, resources, and needs. We must find out how these conditions affect our system. We should continuously analyze the characteristics of our environment, its needs, requirements, the resources available to us, and the constraints that might limit our operations.

Second, we must maintain a high degree of sensitivity toward our environment in order to be able to render an accurate judgment on the relevancy of the input and its value to the system. We need to feed this information back to the environment in order to improve the ability of the environment to accommodate the needs and requirements of the system. A well-informed environment will be much more likely to accommodate its system. A well-informed system will be much more likely to satisfy its environment.

Third, we must ensure that the system will be continuously informed about the adequacy of the product of the system. This requires us to sharpen our ability to measure the adequacy and relevancy of the output.

We have discussed some aspects of systems-environment relations concerning the tendency to isolate a system from its environment. In the application section that follows, you will have an opportunity to review all the main aspects of the systems-environment or school-society models.

Application

In the preceding sections of this chapter you have worked with concepts and principles that constitute the systems-environment model. You have transformed a general systems-environmental model into the specific system space of education, thus creating your own school-society model. These exercises have put you on the road to the acquisition of the systems view—at least in the realm of systems-environment relations.

Working with this application section will help you move further in the systems view. You are now asked to select a specific system, preferably one in which you are involved. Then you will examine this system from the point of view of the system-environment model we have already developed. This examination will take two directions. Using the systems-environment model as a frame of reference, you will be asked to answer a set of questions that will attempt to determine how closely the real system you have selected conforms to the concepts and principles of the systems-environmental model. Then, we will move in the other direction, comparing the systems-environment model to the real system. During the course of your examination, attempt to find the answers to the following general questions: Is the theory implied by the model an adequate one? Does the "map" match the "territory" it is supposed to represent?

By assessing the degree to which the real system conforms to the systems-environment model, you will be able to identify areas of discrepancy between what the model suggests that the system should be and what your analysis indicates that the system actually is. On the basis of this analysis, you will later select one of these areas as the target area for design or for redesign. In the final section of this chapter you will have an opportunity to evaluate the adequacy of the systems-environment model as compared to the real system and you will be encourage to make revisions in the model itself.

The following list of questions will help you explore the system you have selected. The questions are representative rather than all-inclusive, and, as you work your way through them, you might find it useful to add to them or modify them. Their sequence follows, to some degree, the sequence of presentation in the description of the systems-environment model.

1. How is the environment of the selected system defined? Is it described or otherwise documented? Is it a functional suprasystem of the system in question or only a vaguely identified environment surrounding the system?

2. What constitutes the boundaries of the system? To what degree of specificity are the boundaries described? What boundaries are not well defined?

3. Are all environmental entities that may affect the system or that might be affected by the system identified? If not, what appear to be unidentified entities?

4. Are the relationship and interaction patterns between the system and its environment well defined and described? How? What deficiencies appear to be in evidence?

5. What constitutes the input entities of the system? How well are these input entities accounted for? Are there some unidentified aspects of the input? Does the input, for example, include requirements, environmental demands and expectations, constraints, resources, information, and other pertinent details?

6. How well is the output defined? Is it inclusive enough? Does it satisfy the expectations of the environment?

7. How well defined are the ordering relationships of the various systems involved? Are the relationships among the suprasystem, system, peer systems, and subsystems clearly stated? Is the nature of these relationships well defined? (Can they be identified as subordinate, centralized, or equalitarian?)

8. What definitions, descriptions, or evidences are available to define the degree to which the system is open or closed? How varied is the input? How well are the breaks in the boundaries identified? How well are the breaks guarded?

9. To what degree is the system able to adjust to its environment? To what degree is it compatible with it? How does it respond to environmental needs?

10. How is information fed back into the system from the environment? What attention is paid to this feedback? How is it used?

11. To what degree does the environment accommodate the system? How does it respond to system needs?

12. What is your overall assessment? How close is the comparison between your system and the systems-environment model?

Exploring the Adequacy of the Systems-Environment Model

The adequacy of the model can be tested in two ways. First, we can compare the concepts and principles that constitute the model to a real situation to determine if they are adequate. Second, while we are doing this testing, we can identify concepts and principles that are not yet part of the model. To test the model for adequacy, we must consider such aspects as

□ Systems entities in general
□ The attributes and states of these entities
□ Relationships among entities
□ The characteristics of these relationships
□ Processes in which entities engage

Entities of the systems-environment scene were identified in the description of the model as one set of concepts. These entities included (1) the space in which systems exist, (2) the environment (or suprasystem) and its various component entities of which the system is a part, (3) systems and (4) their subsystems, (5) boundaries, (6) breaks in the boundaries, through which (7) input entities enter the system and (8) output entities leave the system.

We can now consider the following questions: In your analysis of the real system, did you discover entities that we did not account for in the model? If you did find some, can you make some generalizations about them and thus complement the model? On the other hand, did you find some entity concepts described in the general model that did not fit into the reality scheme of your system? Report your findings.

Attributes and states of entities comprise a larger set of concepts and principles. They can be described as open, closed, controlled, self-regulating, adaptive, ordered, variant, invariant, random, vague, well defined, stable, unstable, sharply drawn, isolating, or in any other appropriate way. You may enlarge this inventory of attributes from the findings of your earlier examination of a real system in Section 2-2. You might, on the other hand, question the attributes that have been assigned to certain entities in the description of the general-environment model.

Relationships emerge from the interaction of entities. Terms denoting principles that describe relationships in the systems-environment domain include, for example: interrelatedness, adjustment, accommodation, dependence, independence, interdependence, feedback, cause and effect, and trade. *Characteristics of the relationships* were described by such adjectives as hierarchical, centralized, equalitarian, dominating, responsive, and sensitive. Did you find these relationships and their characteristics well exhibited in the real system you analyzed? What were some discrepencies? Did you discover relationships (and characteristics) that were not displayed in the model? How can you include these in the description of the general model?

Processes of the systems-environment scene include interaction, communication, correlation, and control. Some of the terms used earlier recur here in a "process sense," such as *input, output, trade, adjustment, accommodation*, and *feedback*. Most likely you will be able to add to this list some other processes that you identified as you analyzed your system. Keep in mind, however, that only processes pertinent to systems-environment interaction should be considered here.

When you have explored the adequacy of the model by comparing it to an empirical real system of your choice, you can use the findings of this exploration for the purpose of revising or rewriting the general model. The result of this rewriting will be the creation of your own general model of systems-environment. Creating a systems model of one's own is a giant step toward the acquisition of the systems view.

THE STILL-PICTURE MODEL

3

This chapter will describe the principles relevant to the existence of a system within its boundaries. We will develop these principles into a model focusing on what the system is. This chapter will also present some generalizations about systems involving their goals, functions, components, and the relationships among their components. As a result of our exploration of these generalizations, a new *spatial structural systems model* will emerge. This model will display the likeness of a system, its content, and the organization and structure of its content. It will not explain how a system behaves over a period of time, but only what it is at a given moment. For this reason I have labeled this model the *still-picture model*. (In contrast to the still-picture model, a motion-picture model of systems will be presented in Chapter 4.)

In Section 3-1 a general still-picture model will be presented, and you will be asked to transform it into one that represents the school. In Section 3-2 I will introduce my version of the still-picture model of schooling, leaving room for you to develop a synthesis of your model and my version. In Section 3-3 we will attempt to apply the still-picture model to an existing system by means of a structured analysis of that system.

SECTION 3-1   THE GENERAL STILL-PICTURE MODEL

A system embodies three main features: goals, functions, and components. Systems exist for the purpose of achieving certain goals. They employ components that are capable of carrying out the functions required to achieve the goals. The components are "put together" and integrated into a whole to constitute the system in a way that makes for their best use from the standpoint of the system. The goals → functions → components → and their arrangement, then, indicate an obligatory sequence of examination and system building that is most characteristic of the systems view. This sequence will be followed here in presenting the still-picture model. The goal, as the first step in the sequence, also connects the systems-environment model to the still-picture model.

Social systems are goal seeking and goal formulating. They are not only capable of satisfying a goal given by the environment but also of formulating goals for themselves by an analysis of the needs and requirements of the environment.

THE GENERAL STILL-PICTURE MODEL

Goals

YOUR STILL-PICTURE MODEL OF SCHOOLING

Systems come into being or are established in response to a *need* that exists in their environment. It is the perception and assessment of the need that may create a *requirement* for the establishment of a system. (*Need* is defined here as the discrepancy between an existing and desired state, a discrepancy between what something is and what it should be.) In the given space of a system's environment, there might be various needs that compete for attention. Those involved in the creation of a system will consider the competing needs in terms of how valid or critical such needs are. They will also attempt to predict the most likely outcome of attending to one need rather than another. Selection of a need can also be made by keeping in mind the *resources* available to the system and the *constraints* under which it will operate. An analysis of the selected needs, emerging requirements, and available resources will lead to the identification of the *goals* of a system.

To achieve its goals, a system must meet its stated requirements and thus satisfy any specified and assessed need. The goal is the nucleus around which a system grows; a system is built and exists for the purposes of achieving that goal.

The more precise an analysis we make of needs, the more accurately we can define requirements and the more specifically we can state the goal of the system. The more specifically stated the goal, the more precisely measurable is its attainment. The higher the degree of accuracy in measuring the probability of attaining the goal, the more likely it is that the need will be satisfied. At the same time, specifying what the system is doing and how well it is doing it will make it possible for the environment to satisfy the needs of the system.

Thus, a high degree of specification of systems goals is, first of all, *environment serving*, because it makes the system potentially more compatible with its environment.

Second, precise specification is also *self-serving*, because it will encourage the environment to accommodate and support the system more willingly. But specifying goals accurately serves an even more important purpose: the more specific and detailed the goal, the easier it will be to *identify the functions* that have to be accomplished in order to reach the goal. In order to be effective, a system has to activate all functions required for the achievement of its goal. On the other hand, in order to be economical, the system should not engage in any functions that do not contribute to the achievement of the goal.

Functions

There is a clear, logical, and direct relationship between goals and functions as there is also between needs → requirements → goals.

In most statements about systems, the basic importance of functions stands out—*a system is what the system does*; it is a goal-serving function. But what do we mean by the term *function*? And what kinds of functions can we speak of?

There are some functions that are exhibited by every system; we call them *general systems*

YOUR STILL-PICTURE MODEL OF SCHOOLING

functions. And then there are some functions that are necessary for the attainment of the specific goal of a specific system. We call this kind *specific-to-the-system functions*. These designations make no distinction in the importance of the two kinds of functions. In order to attain the goal, both the general functions and the specific-to-the-system functions must be employed.

General system functions denote behaviors that can be observed in any of the kinds of systems we are considering. For example, input and output are functions that are characteristic of all systems. So are growth, feedback, control, adjustment, and accommodation. We have also referred to integration, coaction, and even termination (see Section 2-1).

Another general function is *systemization*. Systemization acts upon the components or parts of the system, brings them together, and makes them increasingly more system-like and goal serving.

The key general function, though, is *transformation*. The transformation of input into output is the most direct goal-serving function. This transformation is often analogous to *production*—the production of an output or outcome that matches the goal, meets the system requirements, and thus satisfies the need. *Information transmission* accompanies transformation, taking the form of information exchange. The transmission of information enables the system to operate and makes it possible for it to monitor itself, thus enhancing its ability to control and adjust itself.

Transformation comes about from activating functions that are specific to the system. Functions that are specific to the system can be identified by carefully analyzing a goal and gradually becoming more specific and detailed. Goal analyzing eventually leads to an identification of functions that the system has to carry out in order to facilitate the accomplishment of system goals.

Specifying functions is important. The more specifically we state functions, the better we are able to identify and select components that have the requisite capabilities.

YOUR STILL-PICTURE MODEL OF SCHOOLING

Components

Parts or components of a system are selected and employed on the basis of their potential to carry out functions to attain the goal of the system. Goals determine the functions in which the system is to engage, and functions determine the capabilities that the system's components should have. *The goal → functions → component sequence is obligatory.* Figure 3-1 displays this sequence.

A system is more than the sum of its components. What makes it so are (1) the *relationships* of the components to each other and (2) the *designed interactions* that fuse these components into a system. It is through *fusing* that components undergo a change and become system-like.

Components of systems are connected by patterned relationships to form the system. These connections are the relationships, and their pattern is the structure of the system. Connections or relationships can be static or dynamic. Static relationships imply unvarying components or subsystems that do not change with time. Static relationships are characteristic of closed systems. Dynamic relationships are characteristic of open systems in which the properties of components or subsystems do vary with time. It is dynamic relationships that we are most interested in, because the systems we are examining are open systems.

It is through dynamic relationships that a

System goals determine ─────────

the *functions* in which the system is to engage.

Carrying out these functions requires specific capabilities that system *components* should have in order to engage in goal-serving functions.

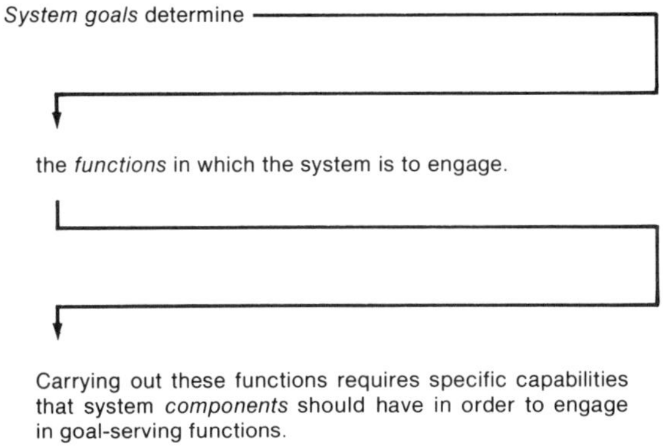

FIGURE 3·1   An Obligatory Sequence: Goal → Functions → Components

system moves from being a collection of independent entities toward the state of integration and interdependence.

Independence of components within a system means lack of integration. It means that
a change in a component or subsystem does
not effect changes in other components. A
tendency toward independence moves the system's components toward *progressive segregation* and isolation and eventually toward the
dissolution or termination of the system. On
the other hand, a *progressive integration* or
systemization within a system moves it toward
a *progressive interdependence* of its components and toward a wholly integrated state.

Interdependence implies that a change in
any part of the system affects all its other parts.
Through progressive interaction, integration,
and interdependence, the system becomes
more than the sum of its parts; this so-called
"synergic quality" is indeed a basic feature
of systems.

If there is true integration, the parts,
components, and even subsystems lose their
original character and become system-like. If
system integration prevails, components take
on the characteristics of the system into which
they are integrated. In the same way, subsystems will take on the characteristics of the
system of which they are a part.

From this discussion, we see the idea of the
wholeness of systems emerging. *Wholeness*
means that although we can think of systems
as being evolved from parts, once the system
has evolved, it is indivisible.

S. Beer notes that division uncouples the
system, interferes with its natural state, and
is, in fact, antisystem. Still, some kind of division is unavoidable, since it is required for the
control of the system. The problem is how to
avoid the point at which division dissolves
the system. "The most profitable control-
system for the parts does not exclude the bank-
ruptcy of the whole."[1] If a division occurs that

---

[1]Beer, S., "Below the Twilight Arch—A Mythology of
Systems," in *Systems Research and Design,* edited by
Donald P. Eckman (New York: John Wiley, 1961),
pp. 16-18.

YOUR STILL-PICTURE MODEL OF SCHOOLING

in the long run affects the integration that makes the system work, then the adequacy, or even excellence, of any given part is irrelevant to the survival of the system. The overriding problem of systems analysis, design, and maintenance, thus, is: How to separate a particular viable system from the rest of the universe without committing an annihilating division.

Summary of the Still-Picture Model

Man-made social systems are created in response to the specific needs and requirements of their environment. The goal of such a system is to meet these requirements and to satisfy the need. The goal of the system determines the functions that are to be activated in order to attain the goal. A system is made up of components with attributes that qualify them to carry out the functions required for the attainment of the goal. Components are transformed and integrated into a whole by the introduction of goal-serving relationships, which determine patterns of interaction and interdependence among components. Components serving specific functions merge into a subsystem. Through progressive systematization, systems become even more integrated, indivisible, and whole. To attain true wholeness is the highest achievement of systems.

The still-picture model can be useful, but it is far from able to present a complete picture of a system. The most obvious limitation of the still-picture model is that it gives us an image of what the system *is*, rather than what the system *does*, and *how* it achieves its goals. Another limitation is that the model is only spatial and momentary and therefore, cannot describe the system over a period of time.

In order to overcome the limitations of the still-picture model, we shall construct another model in Chapter 4. Unlike the still-picture model, it will enable us to include the dimension of *time* in our consideration of systems, as we shall see. But first, let us apply the still-picture model of systems to schooling.

YOUR STILL-PICTURE MODEL OF SCHOOLING

SECTION 3-2 THE STILL-PICTURE MODEL OF SCHOOLING

In this section, on the left-hand side of the page, you will find my transformation of the general model into a still-picture model of schooling. Compare my version with the transformation you developed in Section 3-1, and synthesize the two transformations if you find that such a synthesis would be an improvement. Record your synthesis on the right-hand side of the pages that follow.

THE STILL-PICTURE MODEL OF SCHOOLING

YOUR SYNTHESIS

Goals

School as a system is established by the society to satisfy a great variety of *educational needs*. It is usually beyond the capability of the school to respond to all these emerging needs, even though they might be valid educationally, because society makes limited resources available to the school and often imposes many constraints. Consequently, the school and society must analyze and evaluate what they determine are critical educational needs and assess the probability of satisfying these needs. Thus, by clarifying its educational needs, society will help to establish the *educational requirements* for the school, from which educational goals can then be formulated.

Whether we are within the school system as members or in the general context of society, the more we succeeed in specifying *educational goals*, the more accurately we can measure their attainment. When we know the outcome, we can assess the degree to which specific educational requirements are being met and, hence, the educational needs of the society are being satisfied. Also, clearly stated educational goals will create a more favorable climate in the society toward the school and greater support for it.

Schools should be established around specific goals, and educational programs should be built to meet these goals. The clearer the educational goals are that the school and society arrive at, the greater the probability that the school will see its responsibility to fulfill these goals.

Moreover, if the society knows what the school is doing, it is more likely that it will attend to the needs of the school. We must first formulate expected educational outcomes on specific terms before we can start analyzing them to find out what has to be undertaken to ensure the attainment of those outcomes. The effort of focusing on specific outcomes discourages engaging in activities that will not serve the attainment of those specified goals or outcomes.

Functions

We have derived educational goals from an analysis of educational requirements, and these requirements were formulated on the basis of an analysis of educational needs. The same kind of logic connects educational goals with *goal-serving functions*: functions in which the school will engage are determined by analyzing goals.

In Chapter 2 we identified functions of the school-society interchange. For example, input and output are functions by which education interacts with the society. Furthermore, the adequacy of education is ensured by feedback and control within the school system, making it possible for the educational system to adjust to meet the expected output. Accommodation is another systems-environment interaction in which society fulfills its responsibility to the school by meeting the resource require- ments of the school.

As we saw in our discussion of the general still-picture model, the function operating *within* the system space that is central and crucial to the attainment of the goal is *trans- formation*, the transformation of input into output. The key entity of schooling is the learner, and the main function of schooling is to facilitate the transformation of the learner, from the input state of not yet being able to perform in the desired way to the output state when he can indeeed perform because he has learned. Transformation is thus the most

YOUR SYNTHESIS

salient function of schooling—that is, of a formal education.

Traditionally, transmitting knowledge has been considered the main function of schooling. The main difficulty with this traditional view is that it looks upon the transmission of knowledge primarily as a teaching function rather than considering it from the point of view of the student acting upon what has been transmitted. Probably the most compelling criticism against traditional schooling is that, largely, it only transmits in that way.

Transformation can be studied in terms of *constituent functions* needed to be carried out in order to ensure the transformation of the learner. These functions include making arrangements in the learning environment that will enable the learner to confront and master the learning tasks, activating appropriate conditions of learning, managing these conditions, and helping to assess the progress of the learner. To be effective, schooling has to provide for all functions that are needed for the successful transformation of the learner; to be economical, it should eliminate all the nongoal-serving functions.

Systemization is another central function that enhances the transformation of the learner and thus facilitates the attainment of goals. If systemization works, all entities operating within the system space of schooling are lead to become increasingly more focused on, involved in, and capable of serving the central system goal—learning.

Components

A careful exploration and specification of required functions will enable us to identify who and what are required to carry out these functions, so that the school can acquire personnel and resources.

Schooling is the composite of all its subsystems and their components engaged in the accomplishment of the transformation of the learner, to meet educational goals.

The prevailing practice is to divide schooling into *subsystems*, such as administration,

YOUR SYNTHESIS

instruction, facilitation, and maintenance. Each of these subsystems is comprised of a set of *components* such as human and material resources and parts. Components of the instructional subsystems — students, teachers, classrooms, textbooks, and other materials— are interconnected by a set of relationships. Likewise, the administration and facilitation are established within the system of schooling by *patterned relationships*. Inasmuch as schooling is an open system, relationships on both component and subsystem levels should be dynamic—responsive to changing needs and conditions. It happens only too often, however, that relationships in the school become rigid and static; such relationships then lead toward the closure of the system, toward a lack of sensitivity to external and internal needs, and eventually toward inadequate systems performance.

If the transformation of the learner is the key system function of education, subsystems of schooling should be built around the *instructional* subsystem. This subsystem would then have a centralized relationship to other subsystems. We find, however, that often the relationship within the traditional school systems is hierarchial, rather than centralized, in which *administration* dominates. With the instructional subsystem in a central position, all components of schooling and all its subsystems should become interdependent and integrated in the goal-serving function, which is transformation of the learner. However, the typical prevailing condition in the administration-dominated school is the tendency toward independence of the subsystems and components of schooling—even though integration and interdependence of components may be desired. Often personnel of subsystems in such a school operate without any measurable effect on transforming the learner or enabling the student to learn.

The independence of the different subsystems of schooling in our times has grown to the point where there is often a breakdown of communication between subsystems. And the higher we go in levels of schooling, the more observable is this independence and isolation. If progressive segregation of components prevails within a school system, it may eventually lead to total isolation of parts and a consequent degeneration of the system.

Wholeness is essential in schooling because only through wholeness—or total integration of entities—can undivided attention be paid to facilitating the student's transformation and to ensuring that students will perform as desired. A unique example of wholeness in a system of schooling existed in the one-room school. Of course, the complexity of schooling today does not make it feasible to maintain one-room schools anymore, but it is possible to overcome some of the complexities by paying more attention to the integration of the components of the school system.

Summary of the Still-Picture Model of Schooling

Certain needs of the society create the requirements for educating members of the society. On the basis of those requirements, goals are formulated around which schools are created. By analyzing educational goals, we can identify the functions that the school should engage in to attain its goals.

As a man-made system, education can be deliberately designed for the purpose of facilitating learning. An educational system is composed of human and material components, selected for their specific attributes that enable them to carry out functions to facilitate learning. These components enter into relationships with each other, interact, and become integrated into the total pattern of schooling. A set of components that clusters to carry out a main function may form—for example, the instructional subsystem—and this set of components may assume a coordinating role among other subsystems. A progressive systematization and integration of these subsystems and their components will lead to a school system in which everything and everyone is involved in the facilitation of learning.

YOUR SYNTHESIS

SECTION 3-3 DISCUSSION AND APPLICATION

In this section you will be making use of the understanding of the concepts that define and the principles that govern the way systems are structured and organized within their system space. You will activate the understanding you have developed in working with the still-picture model and move even further toward the acquisition of a systems view.

Discussion

In order to enhance your application experience we shall dispel some of the misconceptions that might block the development of a systems view of structure. A rather typical misconception is to consider the whole as being the sum of invariable parts. This jigsaw-puzzle conception is shown in Figure 3-2. According to this conception, the system can be identified by adding up its parts. The components' parts are fixed, their borders clearly defined, their places clearly mapped out and determined. Parts of this system are stationary, with stable values; the parts are autonomous. This conception of a system is reflected in the way some of our political, social, and educational systems operate.

A somewhat more flexible conception is the push and pull conception (see Figure 3-3). The rigid, fixed relationship is broken up; parts are not stationary, they are energized—they move. They exert energy and force

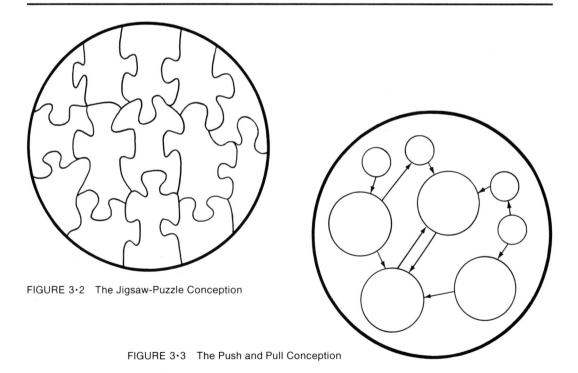

FIGURE 3·2 The Jigsaw-Puzzle Conception

FIGURE 3·3 The Push and Pull Conception

as they push and pull each other. According to this conception, the parts themselves are still invariable and separately defined, even though they are mobile. They engage in functions that are governed by the laws of physics. This conception has characterized the world view of our industrial machine age. According to this view, systems are made up of autonomous entities that attempt to move—or are being moved by—each other toward a goal.

Contrary to the jigsaw-puzzle view, a systems view maintains that the system is more than the sum of its parts. The patterned relationship existing between the system's components and their designed interaction *fuse* them into the unity of the system, so that they lose their autonomy and merge into each other. In this fusion they undergo change; they become system-like as they assume the characteristics of the system.

And in contrast to the push and pull conception, in the systems view the behavior of parts or components is governed by laws that are more like those of psychobiology than physics. The two-dimensional diagram shown in Figure 3-4 can only suggest, rather than depict, the multiplicity of relationships and interaction processes that fuse components into the wholeness of the system.

A tendency to deal with single components rather than with the "whole" is another major hurdle that we must overcome before we can think and act systemically. Considering a part out of the context of its system, dealing with one "thing" at a time, isolating variables, and dividing up the system are some of the ways to violate the wholeness of systems. Systems thinking enables us to handle all parts of a system simultaneously, rather than single parts sequentially. System models allow us to deal with a great number of variables and account for them—not because of what they are in isolation, but because of what they are as a result of their integration into the system and because of what they do jointly.

FIGURE 3·4   The Fused-State Conception

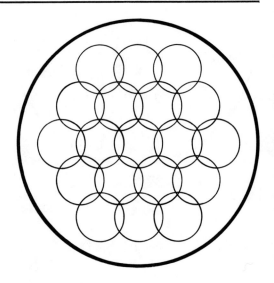

Application

Select a specific system in which you are involved. The system can be the same as the one you have already analyzed in Chapter 2. Now explore this system from the point of view of the still-picture model we developed in the two preceding sections. First, determine the degree to which the real system you selected conforms to the concepts and principles that constitute the still-picture model. Then, test the adequacy of the concepts and principles of the model against the characteristics of your real system.

Examining an Existing System

The questions that follow are representative of the kinds we can ask as we explore the adequacy of an existing system, using the concepts and principles that govern system organization as our criteria. As you consider these questions, you might find it useful to modify them or add to them.

1. Can you determine the sequence that is followed in organizing the system? Was a goal $\rightarrow$ functions $\rightarrow$ components sequence followed?
2. Are the goals of the selected system clear enough? Are the functions operating in the system adequate for the achievement of specified goals?
3. Are all the existing functions goal serving? Are there functions operating that do not serve the goal?
4. Are there some indications of some "missing" functions, which would enhance the attainment of the goal?
5. Are there criteria stated by which to select the parts or components?
6. Are alternatives considered in the selection of components?
7. Do existing components function as well as expected? Are they adequate?
8. What kinds of relationships exist between components? Static? Dynamic?
9. Are relationships hierarchical, centralized, or equalitarian? What are the implications of these relationships?
10. In the first part of this section we explored three kinds of system structure: the jigsaw, push-pull, and fused. Which one of these characterizes the structure of the system you are examining? What are the implications?
11. Does your examination suggest a tendency toward the segregation or integration of components? What are the implications?
12. Do parts or components function independently or interdependently? What are the implications?
13. Summing it up: What evidences exist that speak for the *wholeness* of the system? Or the lack of it?

Use your answers to these questions to judge the adequacy of your selected system.

Testing the Adequacy of the Still-Picture Model

We can test the adequacy of the model by introducing a set of questions, similar to those presented above, that probe into the relevance in actual systems to the concepts and principles that form the still-picture model. In pursuing this new inquiry, keep in mind three kinds of references: the item-by-item description of the concepts and principles of the still-picture model, the thirteen questions presented above in analyzing the real system, and your responses to the questions given and your own questions that arose.

Questions that probe into the usefulness of the model might include the following:

□ How much specifying of a system's goals is actually required to carry out these goals? What determines an adequate amount of detail?

□ Is the goal → functions → components sequence obligatory? Even if we honor this sequence, shouldn't we think ahead and speculate about components, for example, while we are still working on the selection of functions?

□ What governs relationships? Is a hierarchial relationship bad because it is hierarchical? Isn't a centralized relationship to some extent hierarchical?

□ What are the implications of totally fused, totally integrated, and interdependent states of system components?

□ What are the implications of components losing their identify and initial characteristics in order to become absorbed into the wholeness of the system? What does this imply for the uniqueness of an individual who is a system component?

The questions we have explored here are representative; there are many more you can add. You should come up with questions of your own as you test the adequacy of the model.

When you have completed your exploration, you can use your findings in revising or rewriting the still-picture model developed in Sections 3-1 and 3-2. This involvement will lead you to a point where you have created a still-picture model of your own, furthering your acquisition of a systems view.

THE
MOTION-PICTURE
MODEL

4

The systems-environment model introduced concepts and principles relevant to systems in the larger space of their environment and explained the environmental relationships of systems. The still-picture model explained what systems look like; it represented a system in a *static* state. But real systems are not static, they are ongoing and cannot be fully understood from an examination of stationary models only. Static system models cannot exhibit the ongoing nature of systems. To truly understand systems, we need to examine their behavior and observe the change that takes place in their entities, their attributes, and their relationships over a period of time. Thus, we need a *motion-picture model* of systems. By developing and projecting this motion-picture model, I am by no means proposing a model conflicting with the still-picture model. I am, merely adding the *temporal behavioral dimension* to the already existing spatial dimension.

In a manner similar to the description of the two earlier models, we will present a motion-picture model, in Section 4-1, that is generic to social systems; you are invited to transform the model into the context of education. In Section 4-2, we will elaborate on this version of the motion-picture model of schooling, and at the same time compare the model you developed in Section 4-1 with my version. In Section 4-3, you will be able to test the adequacy of the various models and engage in a systems analysis of a specific aspect of schooling that you will select.

35

SECTION 4-1   THE GENERAL MOTION-PICTURE MODEL

A still picture displays a visual image at a given moment of existence. A motion picture is basically a set of still pictures, taken and projected through time, which reveals movement and behavior. The still-picture model of systems has helped us to understand what systems are, and the motion-picture model will assist us to see how systems behave, how they operate.

In considering the behavior and workings of systems, we will first take an overall view in which we will present a model of systems operations in a broad framework. Then we will describe specific systems operations in some detail. Your task is to transform my description of these operations into the context of schooling.

THE GENERAL MOTION-PICTURE MODEL

YOUR TRANSFORMATION

An Overview of Systems Operations

From the systems-environment model and the still-picture model, we have learned that the systems with which we are concerned are rather open systems, brought about by their environment in order to attain certain outcomes. These systems exist to turn inputs coming from their environment into specified outputs. And the systems operate in a way that can correct for differences (through feedback) between the desired output and actual output: they are capable of adjusting themselves to ensure or improve the probability of goal attainment.

With this broad concept of systems in mind, we can see that the operations a system must have include those that process the system input, act upon the input and transform it into the desired state, process the output, and control and adjust the system.

An initial model of general systems operations is shown in Figure 4-1. These operations can be described as follows:

A. *Input processing* implies operations that provide for interaction between the system and its environment or suprasystem, the identification of system-relevant in-

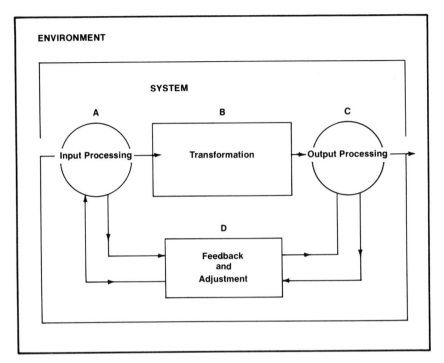

FIGURE 4·1   A Model of General Systems Operations that (A) process the system input, (B) transform the input as desired, (C) process the output, and (D) control and adjust the system.

put, and the introduction of input into the system and the consequent activation of the system.

B. *Transformation* implies operations that bring about conditions by which the input will be transformed into the output.

C. *Output processing* implies operations that provide for the identification and assessment of environment-relevant output, and interaction between the system and its environment to introduce the output into the environment.

D. *Feedback and adjustment* provide for the analysis and interpretation of information relevant to the assessment of the output and, if indicated, the introduction of adjustments in systems operations (A, B, C, and D) in order to bring about a more adequate output. The four major domains of systems operations will now be explored and described in greater detail.

YOUR TRANSFORMATION

YOUR TRANSFORMATION

A MODEL OF INPUT OPERATIONS

Systems are brought about by their environment in order to produce a specific output. The specifications of what this output should be come from the environment and are an important aspect of the input. In fact, the environment is in a prime position in input. It contributes the *subject* of the system, the means of transformation, and the limitations that constrain the system. The subject is the part of the input that has to be transformed by the system from an input state to a specified output state. The resources and energies coming from the environment are the means needed to transform the subject of the system. The constraints under which the system is to operate are also determined by the environment.

There are three significant and distinct aspects of the input process:

▫ Interaction between the system and its environment
▫ Identification of system-relevant input
▫ Introduction of system-relevant input into the system, bringing about the activation of the system

The mode of interaction between the environment and the kind of systems we are concerned about here is *communication*. Requirements, expectations, constraints, and resources have to be communicated. Communication signals relevant to all this input are constantly coming from the environment. As the system interacts with its environment, it receives, decodes, and registers these signals. Signals are registered and screened for their relevancy to the system. The *potential input entities*, which these signals represent, are identified and their values to the system are qualified and, if possible, quantified. Not all potential input becomes actual input in the system. Only systems-relevant input entities are sent into the system, and it is through this process that the system becomes activated.

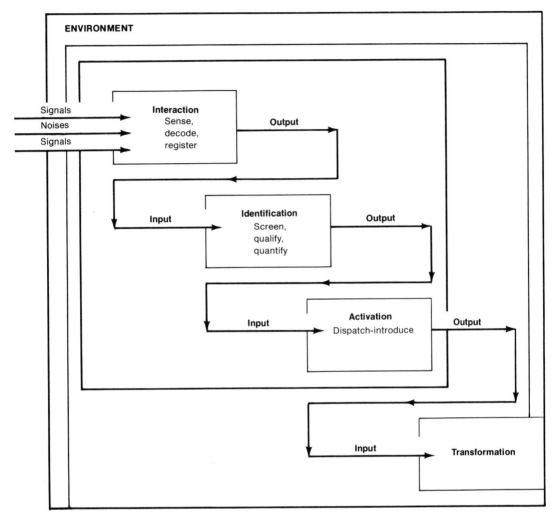

FIGURE 4·2   An Initial Model of Input Operations

These input operations can now be organized into a scheme to represent an initial motion-picture model of the input operations (see Figure 4-2).

Interaction Operations

Interaction embraces the first set of operations input. These operations enable the system to receive, decode, verify, and register signals from the environment. Clear signals are decoded and become specific *input messages*. Other signals may be less distinct or weak, and

YOUR TRANSFORMATION

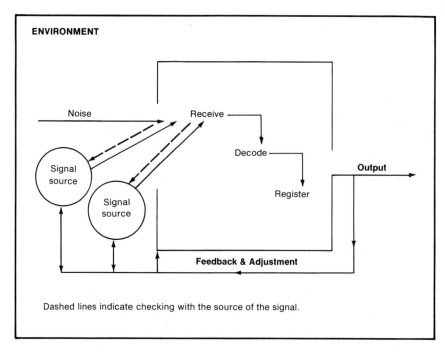

FIGURE 4·3   Interaction Model

---

the message will be less clear. Garbled signals are sensed by the system only as disturbance. If the signal is not clear, the system will activate processes to increase its ability to sense, receive, and decode signals, or it will ask the environment to improve the clarity or strength of the signals. Interaction, thus, also implies that the system will inform the environment—by feedback—of what has been sensed, perceived, and registered. The process of input introduces entities and elements that are intended to be part of the system. A model showing the interaction operations is displayed in Figure 4-3.

Identification Operations

The signals and entities received by the interaction operations become input for identification. The purpose of identification is to interpret the incoming input from the point

YOUR TRANSFORMATION

of view of the system, to select those signals and the entities that are relevant to the purpose and operation of the system, and to qualify or quantify the value of the input.

The selection operation is basically one of screening. Selection operates to allow only those things that are relevant to the system to enter it, and to absorb those things that are not, so that they will not disturb the system. Those who are involved in input identification need information that will determine what is relevant to the system and what is not.

The systems-relevant elements of the input become the true inputs that are now identified according to their value to the system. Depending on the nature of the input and on the devices of measurement available, the assessment of value can be qualitative or quantitative or both, and it can be more or less accurate. The outcome of the assessment is the identification of system-relevant input with specified values to the system. The correctness and adequacy of these identifications are ascertained by the feedback process. The operations of identification are shown in Figure 4-4.

Our criterion for the adequacy of input is the prevailing model of the system in question. This model defines the required state and properties of the input. It is this model

YOUR TRANSFORMATION

FIGURE 4·4   Operations of Identification

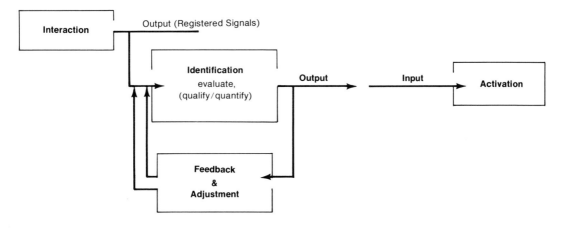

to which we are to compare the outcome of identification in order to introduce adjustments in case there is any discrepancy.

Activation Operations

The last major input operations are the *introduction of the input* into the system and the *activation* of the system when the required input is available. These operations will deliver into the system the specific input needed at the specific place and time. The operations are subject to two kinds of feedback:

- □ The feedback-adjustment of the introduction-activation operations will determine if all input that is needed is delivered and if it is delivered with the required characteristics at the right time and place.
- □ Feedback from the system will also enter as input during activation. This feedback will be admitted into the system to activate needed changes.

Figure 4-5 displays the process of activation.

Summary of Input Operations

An inquiry into how systems work has led us to identify and describe three major input operations that constitute the input process of systems: interaction, identification, and activation (see Figure 4-6).

In our discussion of input operations in this chapter and in earlier systems models, we saw that the outcome of these operations was the introduction of such input entities into the system as:

- □ Those which became the subject of transformation
- □ The system-goal, which specified the desired state of the output
- □ Information about the environment
- □ Resources used to accomplish the transformation
- □ The constraints under which the system is to operate

YOUR TRANSFORMATION

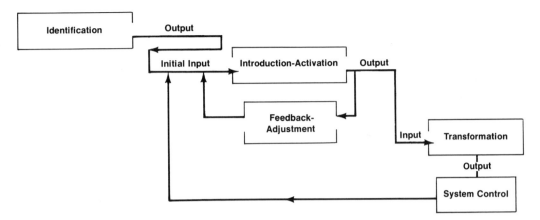

FIGURE 4·5   The Process of Activation

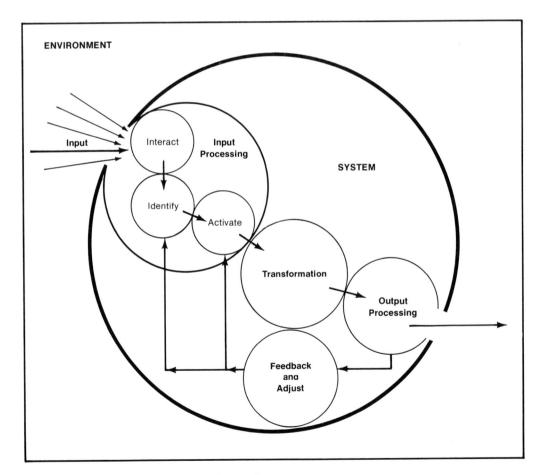

FIGURE 4·6   A Systems Model of Input Processing

A MODEL OF TRANSFORMATION

The four major clusters of systems operations are input processing, transformation, output processing, and feedback and adjustment.

Transformation aims at the attainment of an output. Transformation has three major interacting and interdependent domains (see Figure 4-7).

A. *Transformation production*—in which the subject of the system and other system components are engaged, interact, and use resources in order to accomplish the purposed transformation.
B. *Transformation facilitation*—which aims at the continuous energizing and maintenance of all components that participate in transformation.

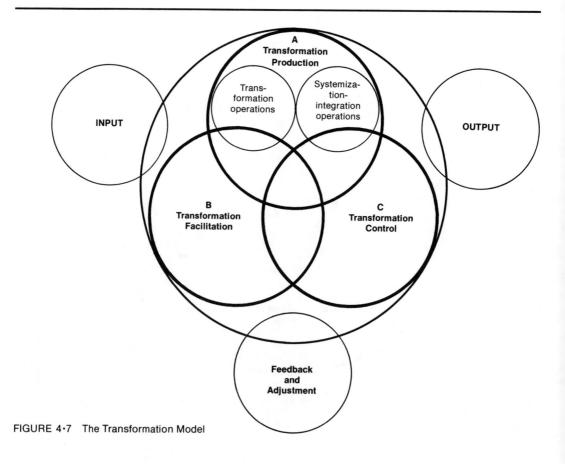

FIGURE 4·7   The Transformation Model

C. *Transformation—control and adjustment* —whereby system transformation is monitored, and the information collected through monitoring is analyzed and evaluated in order to introduce adjustments to optimize transformation.

These three domains of transformation will be discussed next.

YOUR TRANSFORMATION

Transformation Production

Transformation production is composed of operations that aim at the transformation of the system subject from the input state to the desired output state. The output model describes all the aimed-for properties and desirable characteristics of the subject of the system.

Operations that relate directly to this transformation are called *specific transformation production operations.* These operations are designed processes in which the subject of the system and other components are engaged in order to bring about the desired output state.

A second set of transformation operations is called *systemization-integration operations.* These operations aim to make all components of the system more system-like and integrated in order to bring about a state of "wholeness."

At the beginning, a system is set in motion by the purpose and goal around which the system grows and in response to which the system gradually becomes a unified whole. In this process, even the purpose and goal become ever more clarified and specified. During systemization, patterns of interactions among components become ever more specified, weaving them into an integrated whole. Through *progressive integration and systemization,* system components take upon themselves increasingly more the characteristics of the total system, and in so doing they lose their original input characteristics and initial independence as they become ever more involved in the system.

The two kinds of operations of transformation production, the transformation of the subject operations and systemization-integration operations, are simultaneous and interlocked (see Figure 4-7).

Transformation Facilitation

Systems are ongoing, behaving entities. They are involved in transforming input into specified output. In order to be able to do this, (1) systems must maintain behavior that will result in the intended output, and (2) systems need to be energized for such behavior. Operations related to the maintenance of adequate system behavior appropriate to the goal and the energizing of the system components constitute *transformation facilitation*.

Only those system components will be energized that maintain behavior appropriate to the system's goal. This appropriate-to-goal behavior is kept in mind as system components are selected and introduced into the system. Their capability in carrying out the needed transformation operations is carefully assessed or measured. Energizing operations charge up the system components and provide the energy needed to carry out the transformation operations.

Without energizing, we cannot expect effective system behavior. On the other hand, only behavior that will serve the system's objectives should be energized. Thus, through energizing, the system components are maintained in a state that allows them to carry out goal-serving transformation operations.

But how do we know what is appropriate and adequate behavior? Operations relevant to control and adjustment, discussed in the following section, provide us with answers to this question.

Transformation Control and Adjustment

Transformation control is similar to the total systems control (feedback-adjustment) that we have already discussed briefly. However, there are differences in the level of operations and relationships between these two types of control. Systems control operates both within the system space and outside of it. Some of

YOUR TRANSFORMATION

the data upon which systems control is to operate come from the environment or from the suprasystem. The overriding concern is whether the system produces in a way to satisfy the needs, demands, constraints, and broader purposes of the environment. Systems control is made up of an elaborate set of operations that produce feedback to be acted upon by the transformation subsystem or the input or output subsystems. Feedback from systems control may point toward a need to adjust either one of these subsystems or even the feedback control subsystem itself.

Transformation control means (1) monitoring, (2) analysis, and (3) adjustment. Monitoring collects data on all of the ongoing operations of transformation and on the behavior of systems components. These data are analyzed to determine if any discrepancy exists between the intended and actual states of transformation. If there are, needed adjustments are introduced to correct the transformation operations.

The first and main concern of transformation control is to ensure that components of the system perform steadily in a way that guarantees the eventual attainment of the output objectives. Thus, at any point—and systematically at certain intermediate points—performance measures are taken and compared with relevant intermediate performance models. If there is a difference between what the performance of the system is and what it should be, adjustments are introduced to overcome deficiencies. Monitoring, measuring performance, analyzing the data, and adjusting in order to correct for the deficiencies constitute the *performance-effectiveness dimension* of transformation control.

The second concern of transformation control is to ensure that the desired performance will be achieved within the cost-constraints, which is one of the input entities. We will call this concern the *cost dimension.*

Within these transformation control concerns there is a conflict: *cost and effectiveness are usually in opposition.* Often it is impossible for us to employ a component, even

YOUR TRANSFORMATION

though it ensures desirable higher effectiveness, simply because we cannot afford it.

There are occasions, however, when the emphasis may be on the attainment of precision, or at least a high degree of performance, and consequently the cost factor may be of lesser significance. The balance between cost and effectiveness needs to be clearly stated as an aspect of input. Furthermore, this balance is a major factor in deciding whether adjustments should be made.

Adjustments are aimed toward the improvement of transformation operations. To accomplish this aim, we introduce changes that ensure an increased performance of those engaged in the transformation operations, a better maintenance of their capabilities, an increase in energizing and efficiency, and an improvement of the cost-effectiveness balance of the system.

Adjustments are implemented by (1) replacing an existing function or component or interaction pattern with another one; (2) providing for additional functions, components, or interactions; (3) altering or increasing existing capabilities, functions, and interactions; (4) engaging already available capabilities (components) still unused or potentially existing functions and interaction patterns not yet activated; (5) and considering and instituting changes in performance, cost, and effectiveness criteria and thus changing standards by which these aspects are measured.

Transformation control, which is a subsystem of the transformation system, oversees the operations of the transformation production and facilitation subsystems. Jointly, these three subsystems are directly involved in transforming the input into the desired output state (see Figure 4-8).

A MODEL OF OUTPUT OPERATIONS

Output operations concern the identification and assessment of systems-relevant output and the dispatching of this output to the environment or to system control and adjustment (see Figure 4-9).

YOUR TRANSFORMATION

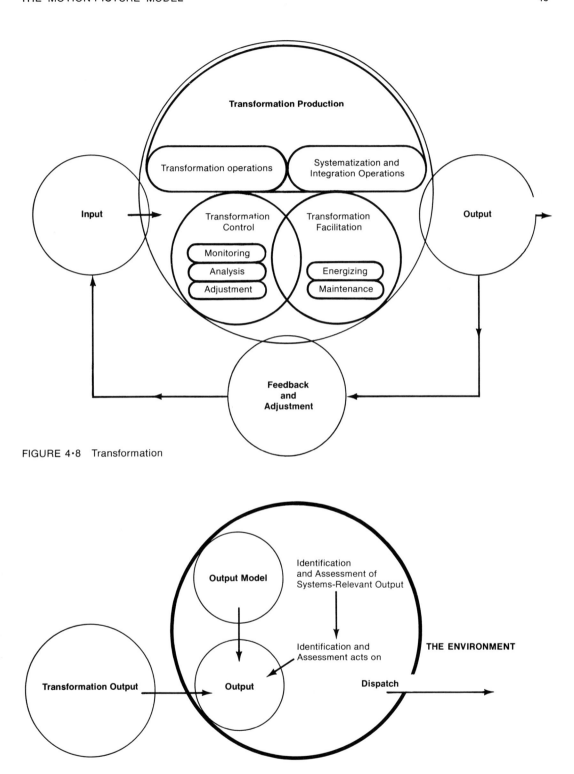

FIGURE 4·8   Transformation

FIGURE 4·9   The Model of Output Operations. The transformation operation and the output model shape the actual output of the system. Output processing assesses the performance of the output before dispatching it to the environment.

YOUR TRANSFORMATION

Identification and Assessment

Upon the completion of transformation operations, we introduce processes that will enable
us to determine if the output meets stated
specifications. Systems may produce outputs
that may not be relevant to the goals of the
system, and systems may produce outputs that
are not adequate—even though they may be
relevant—because they do not match output
expectations. But how do we know this? *How
can we determine relevance and adequacy?*

We need to test and measure the relevance
and adequacy of the actual output, and this
can be done only if there is some model that
represents the desired and expected output
with which we can compare the actual output
to determine the degree of correspondence.
We call this model the *output model* of the
system. The output model helps us to determine the relevance and adequacy of the actual
output. The output model is constructed on
the basis of the goals of the system, with consideration also given to resources available
and specified constraints. Thus, when goals,
resources, and constraints are known, an output model can be constructed and relevant
criterion measures can be developed that can
be used to sample and assess the actual output performance. The adequacy of the output
model and the sensitivity of the measures are
significant, because we cannot detect devia-
tions from output dimensions that have not
been specified or that have not been qualified
or quantified—that is, defined as standards of
expectations.

A comparison between the output model
and the actual output makes it possible to
determine whether or not the output is relevant enough and adequate enough to be
dispatched.

Dispatching the Output

If the comparison between output and output
model shows that the output is relevant and
adequate, the output can be dispatched from
the system.

A MODEL OF FEEDBACK AND ADJUSTMENT

YOUR TRANSFORMATION

The most significant difference between enterprises that operate as systems and those that do not is that systems are controlled by design through feedback and adjustment operations.

System feedback and adjustment comprise the fourth major systems-operations domain in the sequence of input → transformation → output → feedback. Feedback and adjustment operate on evidence collected, relevant to the adequacy of the output and to the functioning of the system (see Figure 4-10).

Feedback and adjustment involve

□ Collecting evidences of the adequacy of the output and systems operations
□ Analyzing and interpreting these evidences
□ Constructing a model of adjustment
□ Stipulating the consequences of adjustment
□ Introducing adjustment in the system

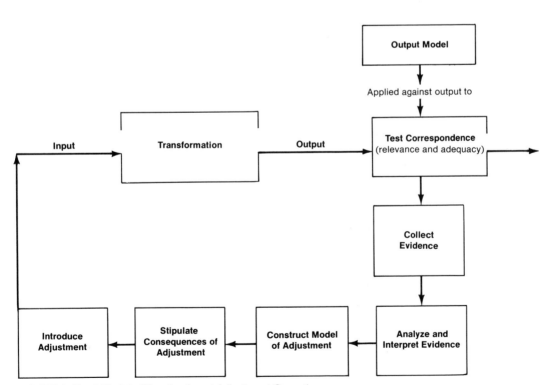

FIGURE 4·10   A Model of Feedback and Adjustment Operations

The information in Figure 4-10 displays the relationship between the operations described here. It also shows us the relationship between output and feedback.

Collecting Evidences

System control is exercised and systems adjustments are made on the basis of an analysis and interpretation of evidences we collect relevant to the adequacy of the output and effectiveness of system operations. Obviously, these two kinds of evidence are interrelated, but in collecting them separately on adequacy and effectiveness we can shift our attention from one to the other.

There are several sources to use in collecting evidence on output adequacy. First, a part of the output operations was to apply the output model against the actual output to see if the output demonstrated the capability projected by the output model and met the performance standards required. If it did, then it was released as a worthwhile product. If not, adjustments were called for. Thus, findings derived from testing the output are our primary source of evidence in assessing the adequacy of the output.

Second, this primary evidence might be augmented by the assessment of the performance of the subject of the system throughout the transformation operations. This assessment gives us insight into the performance history of the subject of the system. The assessment accumulated by the feedback and adjustment operations of the transformation system, while the subject is still being transformed, is often as useful as the testing of the output.

Third, it is the environment that passes final judgment on output. As the system dispatches the output into the environment, an assessment is made on the adequacy of the output from the point of view of the environment. If there is a discrepancy between the assessment of the environment and the system, adjustments are in order. An analysis and interpretation of the discrepancy data indicate

YOUR TRANSFORMATION

the kind and size of adjustments required to correct for the discrepancy.

Evidences relevant to the adequacy and efficiency of systems operations are also collected. These evidences are collected in regard to all systems operations from the initial systems-environment interactions of input operations, through all other operations of input, transformation, output, and also those of feedback and adjustment. Evidences are collected on operational adequacies by comparing the actual performance of the system to the desired operation or to the models of the system operations. This testing of system operations and performance is guided by such overall questions as

□ What aspects are lacking and thus prevent the attainment of the desired outcome?
□ What aspects or components are deficient and thus contribute less than their share to bringing about the desired output?
□ What components produce superfluously?
□ What are those that serve something other than the stated system goals and should therefore be eliminated?
□ Are we getting our money's worth? Can we improve the economy of the system?

Specific inquiries, generated from these broad questions, help us to explore the adequacy of systems components and operations.

Analyzing and Interpreting Evidences

The evidence accumulated from the various sources relevant to the adequacy of the output and the performance of the system are analyzed to determine the nature, the size, and the intensity of the discrepancy between the actual and desired output and the actual and desired operation of the system. The outcomes of these analyses should then be interpreted. The main thrust of this interpretation should be to answer the question: What do the findings of the analyses of the various kinds of evidence mean to the control and adjustment of the system? The analyses will show the dis-

YOUR TRANSFORMATION

crepancy between what the output is and what it should be. Interpretation of the analyses will probe into the source and implications of the discrepancy. Interpretation must answer specific questions, such as: What went wrong? Where, how, and why? Answers to these questions provide the basis of exercising control by introducing adjustments.

Constructing the Model of Adjustment

Once we interpret the meaning of the analysis from the point of view of the system, we can propose certain measures. If introduced into the system as changes, these measures are expected to correct for the difference between the actual and desired states of the output and the actual and desired operation of the system. The organized display or description of these corrective measures constitute the model for adjustment. The interpretation of the evidence revealed the weakness in the system; it told us what caused the less-than-desirable output. We now want to use this valuable information to construct an adjustment model.

The specifics of the model of adjustment will correspond to specific points of system failures or inadequacies. Changes will be designed to correct for these points of failure. According to the nature of system failures, changes will be proposed in functions, interaction patterns, components, distribution and organization, or scheduling. The failure can be of high intensity or low, and the intensity of correction needs to correspond to the intensity of the failure. The key word in designing the adjustment model is *alternatives*. We are to weigh the probable effect of several alternatives and select the alternative that is the most appropriate to ensuring the desired adjustment and correction.

Stipulating Consequences

In a way, the next operation is the testing of alternatives in adjustment. To stipulate consequences is to assess such aspects as (1) the

YOUR TRANSFORMATION

possibility of disruption or disturbance of the system caused by the introduction of the planned change, (2) the constraints and limitations within which the change has to be implemented, (3) cost and time factors, and (4) compatibility with other systems operations and components, and (5) with the environment in which the change will be introduced. These and other relevant factors must be considered in assessing the advantages and disadvantages inherent in introducing the planned change. If one adjustment solution is rejected, another one must be considered.

Introducing Adjustment

Once an adjustment has been tested and a decision made to introduce the adjustment into the system, the appropriate time and the appropriate way to introduce the change must be determined. What is the optimum amount of time needed in order to bring about the adjustment? Can the change be made at random, or should we build into the system by design a specific time to introduce it? To put it another way, will this change be a routine, anticipated operation in the system? Finally, how much time do we need to prepare for and implement this change?

The way in which we introduce change is also significant and should be selected to enhance a clear and unambiguous communication of the change. There must be two-way communication between those who design and introduce the adjustment and those affected by the adjustment. This two-way communication cannot be terminated until feedback is available on the accomplishment of the change and until the desired outcome of adjustments is clearly demonstrated.

Summary of the General Motion-Picture Model

Our examination of systems as stationary phenomena enabled us to develop a still-picture model of systems. When we examined

YOUR TRANSFORMATION

the temporal and process dimensions of systems, we discovered that a new model emerged. This model is a process model, the motion-picture model. Within the framework of this model, subsystems are not formed by any similarity in the attributes of its components, but rather by a similarity of operations in which the components take part. Thus, we find that the model has (1) an input operations subsystem, (2) a transformation subsystem, (3) and output operations subsystem, and (4) feedback and adjustment subsystems. The model that emerged from this examination is depicted in Figure 4-11.

YOUR TRANSFORMATION

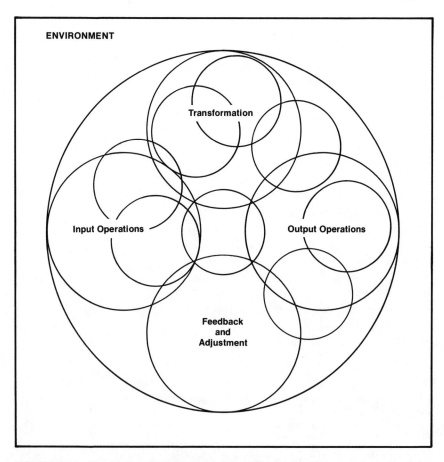

FIGURE 4·11   The Motion-Picture Model. It is difficult to project the true nature of a time-and-motion-determined systems model in a diagram. But, then, how can we expect accurately to depict a motion-picture model in a still picture?

SECTION 4-2 THE TRANSFORMATION OF THE GENERAL MOTION-PICTURE MODEL

The most striking aspects of educational systems are not perceived in what they are, but in what they do. The structural or still-picture systems model developed in Chapter 3 represents a static model of schooling. In this chapter we will construct another systems model of schooling, one that will project what happens in education. We need to depict the movement, the change that takes place over a period of time as the educational process operates, as its components engage in functions, and as the processes are activated that will lead to the attainment of educational goals. This motion-picture model of schooling will add the temporal dimension to the already existing spatial dimensions of our systems models. This temporal dimension will give us the framework within which we can examine what education does and how it operates. The first step toward the construction of the motion-picture model is to construct an overall framework.

On the left-hand side of the pages that follow, I present my transformation version of the motion-picture model of schooling. Compare my transformation with the one you have already developed in Section 4-1, and develop a synthesis of the two versions in the space on the right-hand side of the page.

TRANSFORMATION OF THE MOTION-PICTURE MODEL OF EDUCATION

Schools are brought about by the society in which they operate in order to facilitate learning. If specified learning is not attained, schools are expected to correct for differences between what has been attained and what is expected to be attained. But how does this happen? How do schools operate in order to achieve their goals?

The main operations of schooling include input processing, transformation of the learner into the specified desired state, output processing, and operations that control and adjust schooling. The relationships of these operations are displayed in an *initial model* (see Figure 4-12).

A. *Input processing* include interactions between the school and society. Through

YOUR SYNTHESIS

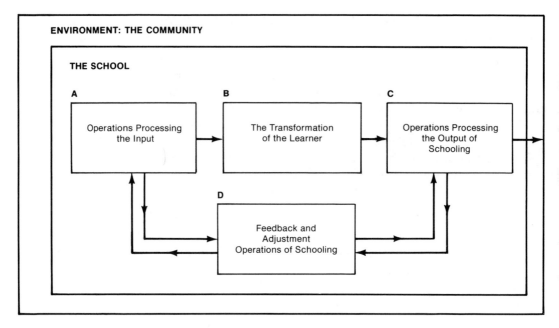

FIGURE 4·12   An Initial Model of the Systems Operations of Schooling

these interactions, the learner and the resources needed to transform him into the desired state are identified and introduced into the system.

B. *Transformation* ensures the availability of means and conditions needed in order for learning to take place. It introduces operations that enable the learner to master the learning.

C. *Output operations* identify the learner who has mastered the learning and assess the degree to which educational goals have been attained. These operations also commission the learner for new learning or for assuming new roles in the society.

D. *Feedback and adjustment* analyze and interpret data acquired from the assessment of the performance of the learner and the performance of the school. On the basis of this analysis, schooling is adjusted to ensure more adequate learner performance and a more effective performance by the school.

YOUR SYNTHESIS

A MODEL OF INPUT OPERATIONS

Education is an open system and its output expectations are determined by the society.

The learner is sent to the school by the society along with other necessary inputs, such as goals, resources, and constraints. The responsibility of the school is to use these input entities in a way that will ensure the attainment of the specified learning.

There are three major classes of input operations through which input is processed by the school. First are interaction operations between the school and its environment, the community or society. Second, the school has to determine which of the numerous input entities and signals received from the society are relevant to the purpose and operations of schooling and to what extent they are relevant. Third, relevant inputs have to be put into the system to activate schooling. Let us now consider these major operations in some detail. (For a map of general input processes, see Figure 4-2.)

Interaction Operations

The school and the community or the larger society continuously exchange information about such things as requirements, expectations, and resources needed by the school. The school *receives, decodes,* and *registers* signals about all of these operations. Some signals are clear and concise; others are vague and appear only as noises and disturbances. The better the school can decode the input signals it receives, the clearer its understanding of the desires, demands, and constraints of the society will be and the better it will appraise the input. As the school receives these signals, it has to feed back to society the results of its decoding to see if the interpretation is correct. When the decoding has been verified, the information must be registered and sent on for further input processing.

The interaction operations of the school can be enhanced by increasing its signal-sensory capability; providing for a continuous surveillance and assessment of educational needs, expectations, and requirements; and involving the community—by design—in the affairs of the school and the school in the affairs of the community.

Identification Operations

The first step in interpreting a signal is to determine if it is relevant to the purpose and function of schooling. The next thing to do is to establish its value—that is, the degree to which the signal is relevant and needed.

Therefore, the recorded input must first pass through a screen that serves two purposes; it has to let through all input that is relevant and, at the same time, it has to absorb everything that is not. Screening protects the school from irrelevant disturbances.

A good screen will ensure that the energies generated (and entered into the systems) for the transformation of the learner operations will not be wasted in coping with phenomena that are not purpose-serving and which should have been screened out during the input operations.

When the input has been identified, a specific value can be attached to each input entity. The importance of attaching the correct value to each entity cannot be stressed too much.

The only way to determine if an input entity sent from the environment into the school is relevant and to assess its value is to compare it to an input model that has been created for the purpose of determining relevance and value. Without such a model, we have no rational basis for coping with input entities.

Activation Operations

Without input there is no production. Once the input of schooling has been received and identified, it must be put to work. The learner must be placed in the learning environment and all the resources—people, materials, and facilities—needed to facilitate the learner's transformation must be activated.

Activation is controlled and adjusted on the basis of two kinds of feedback. First, according to the design plan of schooling, input must be introduced at the time and place required. The adequacy of this process is ensured again by the process of feedback and adjustment, which monitors the introduction of input into the school and brings about needed changes. Second, another type of feedback that also enters here comes from the feedback and adjustment operation, based on the analysis of the output of the school. This feedback carries information and directions back into the system to introduce and activate changes in the input if any are needed, and thus correct for differences between the actual and desired input performance of the school.

Summary of Input Operations

We have identified three sets of input operations by which schooling processes its input —interaction, identification, and activation. Figure 4-13 displays these sets, their relationship to each other, and the overall relationship of input to the rest of the major systems operations of schooling.

A MODEL OF TRANSFORMATION

Transformation operations function to enable the learner to achieve his learning goals. The three operations clusters or subsystems of transformation are (1) *the transformation of the learner operations,* in which the learner and other components of the school engage in order to attain the desired output state, where the learner can perform as expected; (2) *transformation facilitation,* whereby all components of the transformation event, including the learner, are energized and are maintained at an adequate state of performance; (3) *transformation control and adjustment,* by which the transformation event and the outcome of the transformation event (learning) are assessed and interpreted in order to introduce changes

YOUR SYNTHESIS

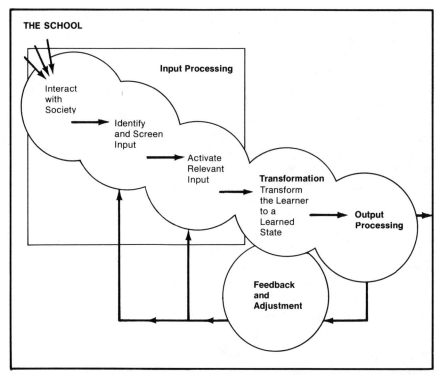

FIGURE 4·13   An Input Model of Schooling Within the Framework of All Other Systems Operations. The four major clusters of systems operations of schooling are shown: input, transformation, output, and feedback.

if needed for the improvement of transformation. Let us take a closer look at these three sets of operations.

The Transformation-of-the-Learner Operations

Two general kinds of operations characterize the transformation of the learner: operations that transform the learner into a "having attained learning" state, and operations that enable the components participating in the transformation to become increasingly more systems-like and integrated.

The first kind of transformation-of-the-learner operations involve:

□ The specification of learning objectives and ways to measure their attainment
□ The identification of learning tasks required for achieving objectives

YOUR SYNTHESIS

- The selection and organization of content that represents the various learning tasks
- The selection, organization, and implementation of learning experiences through which the content is conveyed to the learner
- The management and assessment of learning

The second kind of transformation-of-the-learner operations relate to the systemization and integration of all those who participate in the transformation event and, thus, become increasingly an integral part of the school. In this process, the purpose of learning becomes clearer and more detailed, as do patterns of interactions among such components as the learner, teacher, other personnel, media, and material; and the operations of the transformation become more functional and refined. It is through the processes of integration and systemization that the learner, the teacher, and other personnel become increasingly more involved in the goal and identified with facilitating its attainment.

The two kinds of transformation-of-the-learner operations—operations by which the learner attains his learning goal and the systemization-integration of the components participating in the event—are *simultaneous and interlocked.*

Transformation Facilitation

Components involved in the transformation of the learner need to be maintained in a state that is appropriate to their function. They are to be energized so that they can carry out the functions in which they are involved.

The questions to pursue in considering the facilitation of learning are:

- Does the specific component possess characteristics that enable system-like behavior?
- Does the component make the contribution required to achieve learning?
- Is the component adequately energized toward this achievement?

YOUR SYNTHESIS

If the answers to the first two questions are yes, we should make sure that the answer to the third question is also yes—that is, we should ensure that the transformation component is indeed energized.

One of the aspects of energizing in a system of schooling is motivation, which encourages both the learner and others to maintain the kind of behavior and performance that will facilitate the attainment of learning objectives.

Transformation Control and Adjustment

Operations related to the control and adjustment of the transformation of the learner or the learning event are similar to those that control the total system of schooling. The difference between the two control operations is in the differences in levels (subsystem level *vs.* system level), and the fact that on the system level we are purposely concerned with the suprasystem—that is, with the society. However, on the transformation subsystem level, our primary concerns are the *adequacy of operations* constituting the transformational-learning event and the adequacy of maintenance and energizing.

Control in the transforming-learning subsystem involves (1) monitoring the transformation-of-the-learner event and the maintenance and energizing operations, (2) analyzing the findings of monitoring and, on the basis of those findings, (3) making the adjustment, if any is needed, of transformation functions, interactions, and components. We can analyze our findings by comparing the actually observed learning experiences and outcomes with the model or blueprint of these experiences and with statements on expected outcomes.

The first function of transformation control is related to the performance of the learner. Transformation control monitors and measures the transformation of the learner and the progress he is making toward the attainment of stated performance goals. If there is a difference between the anticipated progress and actual progress, control intro-

YOUR SYNTHESIS

duces adjustments in the learning environment to aid the learner in mastering his learning task.

Transformation control also monitors, analyzes, and corrects the performance of all components involved in the transformation of the learner, in order to assess continuously their efficiency and effectiveness. Another function of control is to check the cost of the ongoing learning program and to ensure that it will operate within set constraints of the system.

There are various kinds of adjustments that alter or increase existing capabilities, components, resources, functions, and interactions of the transformation-learning event; that replace existing functions, components, resources, and interaction patterns of the event; that provide for additional functions, components, resources, and interactions; and that consider changes in the performance criteria for the learner or in cost-effectiveness criteria of the school.

The control of transformation operations occurs through the continuous monitoring of the transformation-learning event, through the analysis of the data collected from the monitoring of this event, and through the adjustments introduced in order to improve the transformation subsystem so that there will be a greater probability of its producing the desired learning.

YOUR SYNTHESIS

A MODEL OF OUTPUT OPERATIONS OF SCHOOLING

The processing of educational output involves operations that enable us to identify and evaluate the output produced and to dispatch acceptable output.

YOUR SYNTHESIS

Identification and Assessment

A system of schooling may produce outcomes that differ both in kind and in degree from what is expected. For example, the performance of the learner may not always be relevant to educational goals. In order to be able to measure differences, we need an output model with which to compare the output behavior of the learner. In this way, we can test the correspondence of the learner's performance with the output model. This model is constructed by considering educational goals, available resources, and existing constraints within which the school operates. However, we need to be prepared to accept outcomes that we have not planned for, which, however, may be of value. We should be prepared to change the output model so that we can attain these added values *by design*.

Output dimensions must be specified and standards of their attainment established on qualitative or quantitative terms, or both. If the learner's performance meets these standards, we can say he has learned.

Dispatching the Output

If the learner's performance corresponds to the output model, then, as determined by society, he can either progress into a new learning program or he can assume his role in the society in some other way.

YOUR SYNTHESIS

A MODEL OF FEEDBACK AND ADJUSTMENT OF SCHOOLING

Changes in schooling should be brought about by design. Any decision to introduce change into the system should be based on a continuous evaluation of the performance of the learner and the performance of the various components of the school. An interpretation of the data acquired from such evaluation will tell us what adjustments are needed in order to make schooling more effective in terms of improving the performance of the learner and making schooling more economical or efficient as a system. The information gathered in mea-

YOUR SYNTHESIS

suring the performance of the learner and of
the system is of no value unless it is used to
bring about an improvement in learning and
in systems operations. *The true control of
schooling means the use of evaluation and
assessment data for the improvement of learning experiences.*

The processes by which such improvement
can be facilitated include collecting evidences
relevant to the performances of the learner and
the various operations of the school, analyzing
and interpreting these evidences, constructing
a model of adjustment that constitutes a plan
for change in schooling, stipulating the consequences of any planned adjustments, and
introducing changes.

Collecting Evidences

There are three sources we can use in collecting evidences that are relevant to the performance of the learner. The first source is the
assessment of the learner we made during the
output operations. Output processing can be
carried out by applying an output model to
test whether the learner has developed the
ability to perform in the way described in
the goals or objectives. If the learner is able
to perform in the desired way, he may proceed into another learning system. If he cannot exhibit the desired performance, he should
be provided with appropriate learning experiences to enable him to match the output
model. Thus, the information we acquired
from the assessment of the learner is our
primary source of output evidence.

This primary evidence is augmented by a
second source of evidences we have accumulated in monitoring the performance of the
learner throughout the learning-transforma-
tion process. Evidences from this kind of
monitoring are often as useful as those which
we acquire from the assessment of the performance capabilities of the learner at the output
point.

When the learner leaves the school, the
adequacy of his performance is assessed by the

YOUR SYNTHESIS

system into which the individual enters. This system is the third source of evidence.

Evidences should also be collected on the adequacy of the operations that facilitate learning. We should compare the actual operations of schooling to the model of these operations and ask such questions as:

□ What operations are lacking, thereby preventing the successful transformation of the learner?
□ What operations are deficient and therefore contribute less than their share to facilitate learning?
□ What operations serve something other than the attainment of learning?
□ What indicates the need to improve the efficiency and economy of these operations?

Analyzing and Interpreting Evidences

The evidences we collect on the adequacy of the performance of the learner and the system must be analyzed and interpreted in order to determine what needs to be corrected or improved. We must find out which specific operations of what components in a particular subsystem are responsible for the shortcomings of the learner or for the less than efficient operation of the system. This analysis and interpretation will supply us with information that will help to determine what learning experiences need to be changed, what operations or components are to be adjusted, what additional resources are needed, and what relationships should be modified.

Constructing the Model of Adjustment

Once we have identified shortcomings and their causes, we must design a plan for adjustment. The basis for planning for change and adjustment in the school is our interpretation of the cause of any discrepancy between the output model and the actual performance of the learner—and between the systems model and the actual system performance.

Depending on the nature of the failure or

YOUR SYNTHESIS

discrepancy, we will have to change some of the transformation-of-the-learner operations or the components engaged in the facilitation of learning and their interaction. We may also find, for example, that the cause of the short-coming is in the input system or in the lack of resources provided to the school by the society. The model of adjustment should respond to those aspects of schooling where discrepancies and failure have occurred. The intensity of change should correspond to the intensity of discrepancy and failure.

In planning for change, we must consider and test alternatives in order to find out which one is best suited to improve the learning experiences of the student.

Stipulating Consequences

In planning for change, we must estimate the effect of a planned adjustment on the performance of the learner and of other components in the school. This will entail speculating about the possibility of any disruption that a change might cause in the learning environment. We must also consider any potentially undesirable effects this change may have on students, teachers, and other personnel. We should also include constraints and cost and time factors in our consideration. We have to test the effect of various change alternatives and choose the one that will be the most effective to improve learner performance and school operations with the least possible disruption and with the least amount of additional cost.

Introducing Adjustment

When we introduce a change into the process of schooling, we must select an optimum time and we must communicate the adjustment in unambiguous terms to those who will be affected by it. We can anticipate a smooth adjustment and change if the school people are aware of it and are motivated to make the necessary adjustment. It is desirable that adjustment to change be a matter of routine and a welcomed

YOUR SYNTHESIS

expectation by the learner, staff, and community. If attitudes conducive to change are not in evidence, then part of the change strategy is to generate such attitudes.

We cannot assume that a change will take place simply because we have planned for it and introduced it. We must have evidence that the change has actually occurred. Acceptable evidence is the improved performance of the learner or the improved efficiency and economy of schooling.

Summary of the Motion-Picture Model

In order to arrive at an adequate model of educational systems, we should examine what such systems are and how they work.

From our examination, a new and probably unique model of educational systems has emerged. Rather than thinking of the school as being made up of such subsystems as administration, instruction, guidance, and facilities, we now think of it in terms of such subsystems as (1) input, (2) transformation, (3) output, and (4) feedback and adjustment.

In the *conventional mode* of school systems, components of the traditional subsystems play their well-differentiated roles as teachers, administrators, counselors, and so forth. These roles are played according to the rules, unchanged even if students fail to perform as expected.

In the *systems mode*, the transformation of the learner is the central process, and all components are interacting in an integrated fashion, rather than in a segregated one, to facilitate this transformation. The system and its components change and adjust by design if this transformation does not indeed happen as expected.

YOUR SYNTHESIS

SECTION 4-3   DISCUSSION AND APPLICATION

In this section you will put to work your understanding of concepts and principles that govern systems operations and behavior. You will apply the motion-picture model we have developed to analyzing a real system of schooling of your choice. But first, let us dispel some misconceptions that might hinder your application.

Discussion

One typical misconception is the belief that the sequence of strategies presented in the motion-picture model implies a rigid sequence in thinking and action, one step *following* another. This is not the case. As we implement systems operations, we discover that *systems thinking* requires us to move *simultaneously* in such additional dimensions as (1) feedback, (2) feed forward, and (3) lateral projection. At any stage of the system operations, we must relate back to the purpose or the goal of the system (*feedback*) and verify and validate all system components and ongoing operations, using specific goal commitments as criteria of performance. By *feed forward* we mean looking ahead in anticipation of what is to come beyond the next step. Feed forward is a speculating process that enables us to deal with the future on the basis of what is now happening and what is known through feedback. Feed forward, thus, can have an influence on what is being done now and on what should come next. *Lateral projection* is the third dimension of systems thinking and is a consideration given to all the functions and activities that occur simultaneously at any given stage of the operation of the system. Being informed about and giving consideration to all aspects of the system's operations, system components are enabled to act and behave in accordance with all that goes on in the entire system.

The systems view requires dynamic multidimensional thinking. It is indeed a challenge for many of us who are used to a sequential, one-step-at-a-time approach.

One-dimensional thinking also makes it difficult to think in terms of alternatives. In systems thinking the consideration of alternatives is more than casual or occasional. It is a deliberate and consistent attempt to thoroughly explore numerous possible means and ways of using functions, components and their interaction, and adjustment patterns with the purpose of optimizing the performance of system operations.

A disposition against change is probably the most potent block of the development of a systems view. We acquire confidence by repetition and replay, and we look forward to becoming more efficient in doing the same thing over and over again. Permanency, familiarity, and routine are powerful sources of security. This kind of security, however, is alien to systems thinking. A common characteristic of all three system models is the requirement of perpetual change that is implemented as a built-in capability of the system. We have to acquire confidence in change, to learn to like it (not only to accept it), and to look for the change and adjustment that will improve the operation of our system and its product.

Summing up our discussion, the motion-picture model has lead us to propose a new structure of schooling, an operation-based structure. The motion-picture model of schooling is a combination of a new set of subsystems, rather than being composed of such traditional subsystems as administration, instruction, counseling, and facilities. These new subsystems are based on operations that are to be carried out to enhance the mastery of learning tasks; they are input processing, transformation, integration, facilitation, output processing, feedback and adjustment. A teacher, an administrator, a counselor, a learner, and a parent who, according to the traditional model of schooling, found themselves to be components of a particular subsystem belong to more than one subsystem —if not to all of them—in the new design. The motion-picture systems model, if implemented, can lead to far-reaching changes in the way education is organized and operated.

Application

You are now encouraged to use the motion-picture model as a scheme for analyzing a real system of schooling of your choice. This system can be the same as one that you have analyzed previously from the points of view of the earlier systems models. The inquiry this time will be somewhat different from previous inquiries. Use the descriptions of the systems operations as presented in the motion-picture model point by point, and ask yourself this question: What evidences do I have that the operations described in the model are exercised or carried out in the real system?

Make note of operations that are provided for and those that are not. At the same time, collect evidence for the effectiveness of the operations. Write down what you believe the implications of your findings are. As you consider the operations and their interactions one by one, you will also test the adequacy of these operations as defined in the description of the motion-picture model (see Section 4-2). This examination will lead you to

□ Confirm the adequacy of the model
□ Uncover operations that the model has not displayed or described
□ Challenge the relevance of the model to some of the actual operations
□ Adjust the description of some of the operations

Finally, use the findings of your explorations in revising or rewriting the motion-picture model. This may lead you to create a model of your own and advance your progress further toward the acquisition and activation of a systems view.

THE
ACTIVATION OF
THE SYSTEMS VIEW

In the three preceding chapters we used a set of general systems models to develop new systems models of schooling. We then used these new models to analyze the operations of actual educational systems. In your work with these models, you may have been wondering which one was the true representation of schooling as a system? The systems-environment model? The still-picture model? The motion-picture model? Of course, no single model is sufficient in itself. Each one presents certain characteristics but not all of the characteristics of systems. The systems-environment model defines the system space; it determines systems-environment relationships; it tells us where and why the system exists. The still-picture model depicts what the system is, what it consists of, and how it is structured. The motion-picture model reveals what the system does and how it does it. *Only if considered jointly and concurrently do these three models tell us the real story of systems. Only together do they reveal the true nature of systems.* It is this fusion of the three models that we must integrate into our thinking in order to acquire a systems view. Only then are we able to analyze schooling, design and develop educational programs or products, and solve educational problems in a systemic way.

5

The process of working with the three systems models, transforming them, and applying them in the context of education has provided the conditions for acquiring a systems view. You should now be able to think of yourself, of the environment you live in, and of the entities that surround you and that you are part of in a new way—the systems way. You have learned to look at education from a systems point of view. You have even begun to activate the systems view as you have analyzed some existing, operating systems. In this closing chapter you will have an extended opportunity to apply your systems view.

DOMAINS OF APPLICATION

The activation of the systems view and the application of systemic thinking mean engaging in operations that will enable us to develop new systems, or analyze existing systems in order to bring about an improved system state and system behavior, or solve problems by looking at a problem as though it were a system and dealing with it as such.

Although these various applications do have specific strategies of their own, they also have much in common because they are implemented through similar processes. The first is the exploration of an area in which you are interested, looking for needs or problems that invite attention, and the selection of a problem. Second, we should specify the desired outcome, consider or invent alternative ways of reaching it, select the most promising alternative, and then plan the move toward the outcome. Third, we develop and test the system or the solution; and finally, we implement it.

The operations we undertake for each of these processes have some recurring characteristics: we consider or invent a number of potential alternative solutions; each solution depends upon a number of factors; the solutions are usually constrained in certain ways; the potential solutions, the influencing factors, and the constraints have various time frames—specific times when they would be possible or in effect; and finally, all of these characteristics are interrelated. What we are faced with is a multidimensional, goal-oriented system of strategies and approaches which we seek to implement in order to analyze, design, and develop systems or solve problems.

Systems solutions are not brought about at certain sequential junctures. We have learned that repeated or recurring application of the same systems strategies leads to strikingly improved outcomes. The strategies of analysis, design and development, testing, revision, and retesting—organized in cycles—bring about improvements and increasingly better products or solutions.

There are various configurations in which we can organize these strategies of systems design or problem solution. The scheme that is selected and displayed in Figure 5-1 is structured in five stages:

  □ *Exploration stage.* What product or system should be analyzed or what problem needs to be solved?

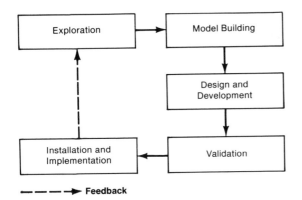

FIGURE 5·1   A Scheme of Activation of the Systems View

□ *Model-building stage.* (1) The construction and testing of alternate conceptual models in order to determine what the product or solution should be able to do and how it should look; (2) The plan of development, validation and installation

□ *Design and development.* The design and the building of a working product or solution

□ *Validation.* By repeatedly testing and revising the working system or solution, we are able to validate it

□ *Installation* and implementation of the system or the solution

Figure 5-1 shows the main stages of the scheme. The arrows indicate the process of continuous evaluation.

The Systems-Models Approach

We can make an important distinction between the *conventional systems approaches* and the *systems-models approach* developed in this book. Most prevailing systems approaches.to analysis, design, or problem solving are sequentially structured procedural steps. They are usually offered without an explanation of how they were developed, how a particular scheme is related to systems concepts and principles, or how it reflects the systems view. Thus, these schemes do not appear to be much more than prescriptions or recipes. The users of these systems approaches are not necessarily knowledgeable about the kinds of systems concepts and principles we have worked with in this book.

On the other hand, *the systems-models approach employs systems strategies, and always keeps in view the systems concepts, principles, and models from which the strategies have been derived.* The systems-models approach is the activation of the systems view that emerges in one's thinking as one internalizes systems concepts, principles, and models. In the rest of this chapter, we will work with this systems-models approach. We will follow the five stages of the system application scheme shown in Figure 5-1, namely (1) exploration, (2) model building, (3) development, (4) validation, and (5) implementation.

EXPLORATION

Exploration leads us to a definition of the system or product that should be developed or analyzed or the problem that needs to be solved. Our conceptual framework of exploration is the systems-environment model. The concepts and principles displayed in this model guide the implementation of the various strategies of exploration. Figure 5-2 depicts these strategies. Strategies of exploration are interactive.

Exploration Strategies

- □ *Define the problem area* in which we are to search for system problems or system needs. Set the boundaries, specify the entities that occupy the problem area, and state their relationships. Examine the problem area for signs and clues that could indicate as yet unsatisfied educational needs or that point to existing educational problems.
- □ *Establish an information and knowledge base* relevant to the defined educational problem area. This is the basis for making solution decisions.
- □ *Analyze the emerged needs or problem signals.* Screen those signals for validity and intensity; establish their criticality. Ask these questions: Is the problem a real one? Is it the kind that can be solved by educational means? Is it of importance? Is its solution feasible within the boundaries of schooling?

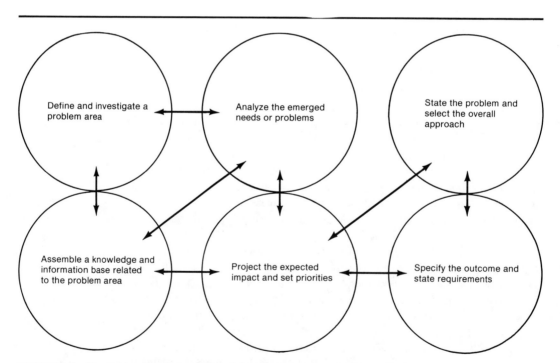

FIGURE 5·2   The Exploration Stage. Strategies of exploration are not linear, but interactive.

□ *Consider the defined needs and problems* as competing against each other for resources. Make a projection of the potential educational impact of attending to these needs or problems. Set priorities based on projections. Select and define the problem or the need. Consider various alternative approaches and, in view of resources and constraints, select the most cost-effective overall approach.

□ *Present an initial specification of the expected outcomes. State the overall requirements* of acquiring the outcomes in terms of resources, cost, time, effectiveness, and efficiency.

The output of exploration is a statement describing the verified need or problem; a rationale for considering the need or problem in terms of how valid, critical, and feasible it is; an analysis of the potential impact of attending to the proposed need or problem; the selected overall approach; and the specification of the expected outcome and systems requirements.

Throughout the exploration stage, our effectiveness depends to a great extent on our ability to conceptualize our involvement on the basis of the laws and principles of the systems-environment model. This model is reflected in a way of thinking that will seek to specify: (1) the boundaries that delimit the system and set it apart from the environment; (2) all the entities in the environment that may affect the system and may be affected by it; (3) the relationship and interaction patterns and the communication channels between the system and its environment; (4) all of the input entities, including the constraints under which the system is to operate; (5) the output that is expected of the system; (6) the order of relationships (suprasystem → system → peer systems → subsystems) and the nature of these relationships; (7) the degree to which the system is closed or open; (8) the degree to which the system can adjust to and is compatible with the environment and the way this adjustment is made; (9) the way the environment can accommodate the system; and (10) the feedback-control process by which the adjustment and accommodation trade off is exercised.

Your challenge is to make use of your newly acquired ability to think systemically about systems-environment relationships. When you begin your systems application project you should

□ Implement the strategies of exploration presented here
□ Guide this implementation by your newly acquired systemic thinking; remind yourself constantly of the systems-environment concepts and principles you have worked with
□ Draw upon the experiences and findings you have accumulated as you have analyzed earlier systems-environment interactions of a real system

MODEL BUILDING

The term *model*, as used here, denotes (1) the conceptual representation of "something to be made," such as a product or solution to be developed.

We can call this representation the product or solution model; (2) *the plan of an operation* that has not yet been accomplished, such as a plan to make the conceptual model real or a plan to design and develop, validate, and install a product or a solution. We will call this the process model.

The outcome of model building is a description or other form of display (such as a mock-up) of what the product or solution should be able to do, how it should look, and a plan or blueprint for how it will be developed. Model building makes it possible for the developer to speculate freely about solutions and to invent various alternatives that can be described and displayed and tested conceptually, in order to arrive at the most promising solution *without* a large investment of resources.

Likewise, building process models gives us a chance to consider various potential alternative approaches to development. The conceptual testing of these alternatives will enable us to select the best possible plan for carrying out a project.

Building Product or Solution Models

A model of a product or problem solution is a written description or some other display of the product or solution. The input to model building is the outcome of exploration; more specifically, the input consists of statements of what product or solution should be developed and the general approach to its production.

An analysis of this input will enable us to formulate a tentative, initial description of what the product or solution is expected to be able to do. This statement will not be a firm description of outcome specifications. Such specifications will emerge gradually as we begin to build product or solution models, as we analyze them for cost effectiveness, and as we select one of the models for implementation. Initially, we need to provide only as much detail about outcome expectations as we need to enable us to invent alternative forms of what the product or solution will be like. We need also to speculate about certain performance indicators that we would accept as evidences of acceptable performance.

The terms *initial, tentative,* and *speculative* have been used here to indicate open rather than closed decision-making and to demonstrate an iterative approach of successive approximation. (A closed state implies clearly defined goals and highly specified resources, constraints, and levels of performance. In a closed state there is little if any interaction allowed with the environment. Openness, on the other hand, indicates a continuous sensitivity to environmental influences—a level of aspiration rather than a precisely defined goal, a gradually unfolding specification of expectations, empiricism, and flexibility.)

A product or solution specification will emerge through repeatedly applying the strategies of analysis, synthesis, evaluation, and revision to performance specifications. The reapplication of these strategies will enable us to eventually reach a point where we can make a firm statement about product or solution specifications.

Once an initial statement of expected outcome is provided, even if it is a tentative one, we will consider or invent alternative models of the product or solution and describe or otherwise display these models. In describing the alternatives, we should also specify the potentials and limitations of the various alternatives.

With a set of models available, we can then test them for cost effectiveness and feasibility. More specifically, the testing of potential models should respond to the following inquiries:

□ What evidence can we provide or suggest that would demonstrate the effectiveness of the various alternative models? Which model would best meet the initially stated performance specifications?

□ If we consider potential resources and existing or anticipated constraints, how feasible are the various alternative models? What would be the expected cost of development or solution and cost of implementation and maintenance?

Testing the models can be done in two ways. One way is to test them conceptually, by thinking through how the models would perform under projected conditions and by making a cost-effectiveness estimate for performance and a feasibility estimate for development. Another way is to conduct a computer simulation, provided the anticipated performance characteristics and key aspects of the performance can be quantified. The outcome of testing is a prediction of both cost effectiveness and feasibility. On this prediction, the best performing models can be selected for implementation.

The output of constructing product or solution models is a statement that describes the requirement for the product or solution; the initial performance specifications with performance indicators also noted; the alternative models of the product or solution; the analysis and testing of alternatives and the outcome of testing, including a statement of why a specific model was selected; a detailed description of the selected model, which in fact is a restated description of what the product or solution will look like and how it will perform.

Building the Process Model

While we are building the models of the product or solution, we should speculate on the various ways of developing the product or solution. We should consider alternative planning strategies, test the alternatives conceptually or through simulation, and select the most cost-effective strategy.

The outcome of building the process model is a description of the specific strategies to be used, including the evaluation method, and the implementation events and activities organized in a scheme with an estimate of time, cost, required personnel, and material resources. In brief, the outcome of model building is a comprehensive description or display of the product or solution and a plan by which to acquire the product or solution.

As you continue the planning and design of your own system, following the strategies of model building already described, you should be guided by the concepts and principles we organized in the frameworks of the still-picture and motion-picture models. The still-picture model is manifested in such tendencies as exhibiting a high degree of specification of the systems goal; insisting upon the sequence of goal → functions → components; clarifying all system functions that are goal serving: eliminating non–goal-serving functions and activating those that do serve the system; considering alternatives and employing components that have the competence to carry out required functions; establishing and maintaining relationship and interaction patterns between components and subsystems in a way that ensures their integration into the total system and that optimizes their contribution toward the goal of the system; and adhering to a trust to maintain the indivisibility and wholeness of the system. The concepts and principles that we generated as we explored the motion-picture model of systems should be interacting with our still-picture model view. To review the points of the motion-picture model: a system is primarily and above all *process*, and what the system does determines what it is; the change that takes place in the attributes of the system through time is important. The four main systems processes are input, transformation, output, and feedback and adjustment. Input is more than a break in the system boundaries through which input enters into the system. It is a set of operations of receiving, decoding, registering, screening, and evaluating the input for its relevancy and value to the system and then admitting (and energizing) the systems-relevant input. Transformation, another set of sequential and interacting operations, enables the system to transform its subject into the desired output state. Output is not only another break in the system boundaries but, more importantly, it is made up of operations that assess the adequacy of the output. If it is adequate, the output is dispatched into the environment; if it is not, the discrepancy is noted and fed back. An adjustment is constructed that will correct for the difference between the existing state and the desired one.

As you continue your system application project and begin to design the model of your new system or the model of the solution of your selected problem, you will:

□ Implement the strategies of model building described here
□ Guide this implementation by the system concepts and principles inherent in the still-picture and motion-picture models
□ Draw upon the experiences you accumulated earlier while you analyzed your real system

Preparation for an Extended Application

In the course of applying the systems framework in this book, it was anticipated that you would complete the exploration and model-building stages of your selected project. By now you have

□ Formulated the specifications of your system, product, or problem solution
□ Selected the most promising product or solution alternative
□ Developed plans for carrying out the design, development, validation, and implementation stages

These three achievements are some of the tangible results of your effort in applying the systems view. These results become the input into an extended application of your newly acquired competence in the use of the systems-models approach.

The plan that you have developed for the design, development, validation, and implementation stages of your project becomes the "contract" that you have made with yourself for such an extended application. This extended application may take several months, as you will actually build, test, and install your system, your product, or your problem solution. In the rest of the chapter we briefly look at the strategies of design, development, validation, and implementation.

The description of these strategies, but even more importantly that of the three systems models and your systems view, will guide you throughout your extended application.

DESIGN, DEVELOPMENT, VALIDATION, AND IMPLEMENTATION

Throughout these stages you will invent initial design solutions, build a first working model, develop measures to assess its adequacy, test, analyze findings, revise, rebuild, test again, analyze findings again, and revise again. You will need to continue these operations until you have accumulated adequate evidence that your product or solution meets stated expectations and until you have developed enough confidence in it to assure yourself that it will work. The number of cycles you will have to go through depends upon a number of factors, such as your competence to work out cost-effective solutions, the complexity of your project, the relevance of your solution to the system in which it will be installed, and the acceptance of it by those involved in the system.

Design

During the model-building phase, you made plans of how to develop your product or solution. Design is basically a detailed elaboration of this plan. It involves the design of the first working form of the product, system, or solution. The design is the bridge between the conceptual model and the "real thing." If your project is more complex, for example, if in your conceptual model you have projected a solution that has several components, then you will need to undertake design at both the "components" and the "whole" levels. You should first make a tentative design at the whole level, which becomes the framework for evolving an optimum configuration of components, then work out the detailed design of components and integrate them in the more refined design of the whole.

Strategies of design include (1) accumulating the information required to develop the design, (2) formulating design alternatives, (3) testing these alternatives against stated product or solution specifications, (4) preparing a design description, and (5) preparing instructions for the development of the first working model, which is often called the *prototype*.

Development and Validation

It is hard to say where design leaves off and development begins; they blend into each other. It is also difficult to separate development from validation. As we have already seen, strategies for these processes are continually repeated until we are ready to implement the model.

When you build the first prototype of your system (or product or problem solution) and test it against specifications, you will develop information that usually leads to the revision of the design and results in more refined design specifications. The revised design becomes your blueprint for building a new improved form of your product or solution. Your main interest in developing your first working model is to find out if your product or solution is workable at all. Your main concern in developing the second or revised form is to make an assessment of how well your product or solution performs against the stated performance specification; thus, we often call this second form the *performance form*.

In order to assess the performance of your product or solution, you will have to have indicators available (or you will have to develop them) that will test the adequacy of the performance. Make sure that these indicators are representative of the expectations of the eventual users of your product or solution.

Your next step is to analyze the findings of your testing of the performance form by applying the strategies of feedback and adjustment of the motion-picture model. On this basis you will be able to identify any discrepancy between the actual and desired performance of your product or solution and develop an adjustment model to use as the basis for revising your product or solution.

Testing and testing-based revision will lead you to a state of validation or quality assurance when you have accumulated enough evidence to support a decision to install your product or solution.

As you proceed through the stages of design, development, and validation, you should be continuously guided by the concepts and principles of the three systems models.

Implementation

After you have developed a tested product or solution, you will want to install or implement it within the framework of a larger existing system. It was your exploration of this larger system that led you to identify the educational need the product was designed to satisfy or to specify the problem you wanted to solve. The strategies of adjustment, described in the motion-picture model, will serve as a guide in implementing your product or solution.

We have now reached the end of the book and a beginning of a new venture. Having worked with this book, you are now ready to commence a continuous, ever-increasing application of the systems view to your day-by-day work. We hope that you will also help others acquire the systems view, and continue to bring about improvements in the field of education.

GLOSSARY

Accommodation is a systems-environment interaction or process by which the environment satisfies the changing requirements of the system.

Activation operations are part of input processing (see below). They introduce into the system the input required to operate the system.

Adaptive systems are capable of adjusting themselves to meet changing requirements.

Adjustment is a systems-environment interaction or process by which the system responds to the changing requirements of its environment.

Adjustments are changes that a system brings about in order to modify its behavior, structure, and characteristics so that it can produce improved system output or system state.

Behavioral system model (see also Process Model) describes the behavior of the system over a period of time and displays concepts and principles relevant to systems operations.

Boundaries of a system delimit the system space and set aside from the environment all those entities that make up the system.

Centralized relationship indicates that a given subsystem plays a central and coordinating role and that other subsystems are related to it and arranged around it.

Closed (system) refers to a state of isolation in which the system is sealed off from its environment by its boundaries. There is no interchange between a closed system and its environment.

Closure by control is an ability to adjust the breaks in the boundaries of a system. This enables the system to regulate input and output.

Components are integral parts of a system, selected on the basis of their potential to carry out functions required for the achievement of the system's goal.

Constraints are known limitations or restrictions imposed upon a system that curtail resources or operations.

Cost effectiveness is a consideration of what degree of system or product effectiveness can be attained for what cost; or, given the cost, how much effectiveness can be attained for it.

Decoding is the extraction of information from signals that are coming from the environment.

Entity is a definable element of a system.

Environment is the context within which a system exists. It is composed of all the things that surround the system, and it includes everything that may affect the system and that may be affected by the system.

Equalitarian relationship indicates equality among subsystems. In this kind of relationship, none of the peer subsystems plays a dominant or central role.

Exploration includes the examination of a problem, the identification of needs or problems that invite attention, the selection and definition of a specific problem, and the formulation of expected outcomes.

Feedback is a process by which information concerning the state of the output and the operation of the system is introduced into a system.

Feedback and adjustment provide for the analysis and interpretation of information about the assessment of the output and the operations of the system. This information is used for introducing adjustments into the system in order to bring about more adequate output and improved system operations.

Feed forward is a process in which we look ahead in anticipation of what is to come beyond the next step. This speculation enables us to deal with the future on the basis of what is happening now and what is known about the past.

Functions are activities that have to be carried out in order to achieve the goal of the system.

General system functions are functions that are characteristic of systems in general.

General systems research identifies elements that are common to systems in general and it develops and tests models that represent systems in general.

General systems theory presents concepts, principles, and models that are common to systems in general and it identifies structural similarities between systems.

Goal-formulating systems are those which are able to analyze systems requirements and evolve their own goals.

Goal seeking is a characteristic of systems by which they are directed toward the achievement of goals.

Hierarchical relationship is one in which one subsystem is superior to others.

Identification operations are part of input processing (see below). They interpret the input from the point of view of the system and qualify and quantify its value.

Independence of components within a system means a lack of integration. A change in one component does not bring about a change in others.

Input includes information, people, energies, and materials that enter into the system from the environment. It is also the process by which such entry occurs.

Input entities are information, people, energies, materials, and other resources that enter the system.

Input model describes in an organized way elements and resources required by the system and the operations by which input is selected and introduced into the system.

Input processing refers to operations that provide for (1) the interaction between the system and its environment, (2) the identification of systems-relevant input, and (3) the introduction of system-relevant input into the system.

Integration operations intend to make the components of a system increasingly more fused and interactive.

Interaction operations are part of the input processing. They enable the system to receive, decode, verify, and register signals and input elements coming from the environment.

Interdependence of components within a system means that change in one component brings about changes in others.

Lateral projection gives consideration to all simultaneous functions and activities of a system.

Model may be (1) a representation or abstraction of a real system or (2) a theoretical projection or display of a possible system.

Model building is strategy by which a conceptual representation of a system or a solution is constructed and from which specified outcomes can be determined.

Model of adjustment is a conceptual representation of measures that can be taken to correct system performance or the system output.

Motion-picture model is a phrase used in this book to illustrate the behavior of systems that takes place over a period of time, and to elaborate on the concepts and principles relevant to such behavior.

Multisystem is a complex of several related systems.

Open refers to a state in which a system is continuously interacting and interchanging with its environment.

Output is whatever the system produces and sends back into its environment.

Output model describes the intended properties and characteristics of the output of the system.

Output processing refers to operations that identify and assess system-relevant output and introduce that output into the environment.

Patterned relationships are connections between the components of a system. These relationships make up the interactive functions that components carry out by design and that display the structure of the system.

Peer systems are related systems that make up a larger system.

Performance model describes the desired performance of a system.

Process model describes the behavior of a system over a period of time. It displays systems concepts and principles relevant to the behavior of systems.

Progressive integration fuses the components of a system into increasingly more wholeness (see below) and unity.

Progressive segregation comes about as components of a system become increasingly more independent and isolated, leading to the eventual dissolution of the system.

Resources are information, people, materials, money or other means that are at the disposal of a system.

Self-regulating systems are able to modify their own behavior in order to enhance the production of the desired output.

Social systems are adaptive and complex systems composed of causally related components. The interrelationship of the components constitutes the structure of social systems and provides for their wholeness.

Specific-to-the-system functions are functions that are found only in a given system.

Still-picture model is a phrase used in this book to illustrate what systems are at a given moment. The model presents the concepts and principles relevant to the existence of systems, their content, organization, and structure.

Structural (spatial) systems model is a scheme that organizes the concepts and principles that define what systems are at a given time and displays their organization and structure.

Subject (of a system) is the entity around which the system is organized and which has to be transformed by the system from an input state to a specified output state.

Subsystem is a component part of a system. It is made up of two or more components. With a goal of its own, it interacts with it peer subsystems in order to achieve the overall goal of the system.

Suprasystem is a system that is made up of a number of component systems.

System is an interacting group of entities forming an organized whole. More specifically, and in the sense used in this book, a system is a deliberately designed synthetic entity made up of diverse but interdependent components that interact and are united according to some organizing idea, plan, or central principle. A system becomes more than the aggregate of its components.

System concept refers to an aspect of systems, such as "input" or "transformation."

System control is a process by which the system regulates itself or by which the behavior of the system is regulated.

System design aims at the construction of a model or a "blueprint" of a system to be developed.

System development involves the formulation, testing, revision, and validation of a system.

System-environment coactions are processes by which the system adjusts to the changing requirements of its environment and the environment accommodates to the changing requirements of the system.

System-environment model examines systems in the context of their environments and organizes and displays concepts and principles relevant to this examination.

System requirements are the specific demands and conditions that the system is to satisfy.

System space is the domain that the system occupies as defined by its boundaries.

Systemization is a transformation process by which components of a system are fused and become increasingly more system-like.

Systemization-integration operations of transformation aim to make the components of a system increasingly more system-like in order to bring about full integration and achieve "wholeness."

Systems analysis can be of two kinds. Exploratory systems analysis aims to lead to the definition or selection of a problem to be solved or system to be designed. Evaluative systems analysis probes into the adequacy of an existing system.

Systems models organize and present in a scheme system concepts and principles.

Systems models approach is the use of systems models to analyze, design, and develop systems or solve problems.

Systems operations are components of the major systems processes of input, transformation, output, and feedback and adjustment.

Systems principles are constructed from related system concepts. They display the laws that regulate and describe systems. For example, the more complex the input, the more complex the system.

Systems research studies the structure, organization, and behavior of systems and it develops and tests generalizations derived from such studies.

Systems strategies are components of the System Model Approach applied to the design and development of systems or products; or to the solution of problems.

Systems theory presents concepts, principles, and models that describe the structure, organization, and behavior of systems.

Systems thinking is thinking that is influenced and guided by systems concepts, principles, and models.

Systems view develops as systems concepts, principles, and models become integrated into one's own thinking.

Trade-off is an exchange of one aspect for another and the consideration of the effects of such exchange.

Transformation is the process by which the input is changed into output.

Transformation control and adjustment are operations whereby transformation is monitored. The information gathered through monitoring is analyzed and interpreted in order to introduce adjustments by which to improve transformation.

Transformation facilitation is a set of operations aimed at the continuous energizing and maintenance of all components that participate in transformation.

Transformation processing refers to such operations as transformation production, systemization, transformation facilitation, and transformation control.

Transformation production is a set of operations in which the subject of the system and other system components are engaged, interact, and use resources in order to accomplish the purposed transformation (of the subject of the system) and thus achieve the goal of the system.

Transformation model is a conceptual representation of operations by which the input is changed into output.

Wholeness (of systems) refers to the integrated, fused state of the components of a system by which the system becomes indivisible.